Political and
Social Posters
of Switzerland

Politische und
soziale Plakate
der Schweiz

Affiches politiques
et sociales
de la Suisse

Willy Rotzler/Karl Wobmann

Political and Social Posters of Switzerland

Politische und soziale Plakate der Schweiz

Affiches politiques et sociales de la Suisse

A historical cross-section

Ein historischer Querschnitt

Un aperçu historique

ABC Edition Zurich

ABC Verlag Zürich

Editions ABC Zurich

© 1985 by ABC Verlag, Zurich
ISBN 3-85504-085-0
Printed in Switzerland

© 1985 by ABC Verlag, zürich
ISBN 3-85504-085-0
Gedruckt in der Schweiz

© 1985 by ABC Verlag, Zurich
ISBN 3-85504-085-0
Imprimé en Suisse

	Contents		**Inhalt**		**Sommaire**
7	The Poster as Envoy	7	Plakate als Botschafter	7	L'affiche, messagère par excellence
16	Insurance for the Elderly and Bereaved	16	Kampf um Alters- und Hinterbliebenen-Versicherung	16	Lutte pour l'assurance-vieillesse et survivants
24	Women's Suffrage	24	Das Frauenstimmrecht	24	Le droit de vote des femmes
40	Shorter Working Hours	40	Kürzere Arbeitszeit	40	La réduction de la durée du travail
52	Atomic Defence	52	Atomschutz	52	La protection contre les dangers atomiques
62	First of May	62	Der 1. Mai	62	Le 1er mai
68	Winter Aid	68	Die Winterhilfe	68	Le «Secours suisse d'hiver»
82	For the Old	82	Für das Alter	82	Pour la vieillesse
94	Aid for Refugees	94	Hilfe für Flüchtlinge	94	L'aide aux réfugiés
98	The Red Cross	98	Das Rote Kreuz	98	La Croix-Rouge
106	Pro Infirmis	106	Pro Infirmis	106	Pro Infirmis
110	Pro Juventute	110	Pro Juventute	110	Pro Juventute
128	Aid for Children	128	Hilfsaktionen für das Kind	128	Actions de secours en faveur de l'enfant
134	International Aid Appeals	134	Internationale Hilfsaktionen	134	Actions de secours internationales
138	Protection of Nature and Wildlife	138	Naturschutz – Tierschutz	138	Protection de la nature – protection des animaux
142	Protection of the Environment	142	Die Umwelt schützen – uns vor Immissionen schützen	142	Protéger l'environnement – se protéger contre les immissions
146	Campaign against Drug Abuse	146	Kampf gegen Rauschgift	146	La lutte contre la drogue

The Poster as Envoy

In the course of more than a hundred years of poster history there have been numerous attempts at classification in order to give an overall view of this abundant and diverse art form. Everyone of these principles of classification—whether by artist, country of origin, period, graphic technique, style, or especially purpose, theme and subject—provides its own insights into poster art. None of these insights, taken alone, provides satisfactory answers to all the questions, so it is particularly useful to combine them.

If one starts with the purpose of the poster, that is its specific advertising and information functions, there is an immediate division of all poster production into the two main areas of commercial and non-commercial subjects. This broad division rests on the assumption that, in the course of historical development, there have been separate "tracks of requirement", scarcely comparable with one another, for which the poster has been, and still is, considered suitable.

The poster can be seen as an invitation, a kind of envoy, made possible by graphic reproduction, widely distributed and addressed to a mass of anonymous individuals, not accessible by other means. The object of this emphatic invitation can be seen as activity, even if this activity only consists of the attempt to satisfy a need—aroused, enhanced or suggested by the poster—for an object, a product or a service.

With the commercial poster, this means that one acquires a consumer article, a product, such as a packet of cigarettes, or that one makes use of a service, for example by obtaining a credit card or a membership in a broadcasting system by the telephone network. The commercial area also includes all the services offered by travel and tourism: one does not simply buy an air ticket but also a holiday on a sunny shore. The "business talk" conducted by the poster with the viewer has become more subtle, one might say more refined, in the course of time. The poster advertising sun-tan lotion does not only sell a protective filter against ultra-violet rays; it also sells a healthy brown skin, and much more: youth, beauty, attractiveness; thirdly it sells holidays, leisure, freedom from daily work and worries: in a word, happiness.

In addition to the poster which offers something commercial there is the wide area of the non-commercial poster. It invites us in an amiable way, for example, to be polite and friendly to one another; it recommends us, with an undertone of admonition, even indirect threat, to stock sufficient emergency supplies "now" for our own and our country's safety; it warns us of the effects of drinking on driving and against pickpockets. In times of emer-

Plakate als Botschafter

Im Verlauf der über hundertjährigen Geschichte des Plakats haben sich verschiedene Methoden eingebürgert, um die Fülle und Vielfalt an Plakaten überschaubar zu machen. Jedes dieser Ordnungsprinzipien – zum Beispiel nach Künstlern, nach Herkunftsländern, nach Entstehungszeiten, nach graphischen Techniken, nach Stilrichtungen und, vor allem, nach Aufgabenbereichen, nach Themen und Motiven – bietet andere Einsichten in die Plakatkunst. Keine dieser Einsichten liefert – für sich genommen – auf alle Fragen befriedigende Antworten; erst ihre Kombination ist ergiebig.

Stellt man die Aufgaben des Plakats, also seine spezifischen Werbe- und Informationsfunktionen, in den Vordergrund, so ergibt sich zunächst eine Aufspaltung der gesamten Plakatproduktion in die beiden Hauptbereiche der kommerziellen und der nichtkommerziellen Plakate. Diese Grobunterscheidung geht davon aus, dass es im Verlauf der geschichtlichen Entwicklung kaum miteinander vergleichbare «Bedürfnisstränge» gegeben hat, zu deren Erfüllung man das Plakat für geeignet erachtete und noch immer erachtet.

Man kann das Plakat als eine Aufforderung verstehen, als eine durch die Mittel drucktechnischer Vervielfältigung potenzierte, weiterum verbreitete, eine Masse anonymer, anders nicht erreichbarer Individuen ansprechende Botschaft. Man kann es verstehen als eine Einladung, eine nachdrückliche Aufforderung zum Tätigwerden, bestehe dieses Tätigwerden nun darin, dass man die – vom Plakat bewusst gemachten, gesteigerten oder suggerierten – Bedürfnisse nach einer Sache, einer Ware oder nach einer Dienstleistung zu befriedigen versucht.

Beim kommerziellen Plakat würde das heissen, dass man ein Konsumgut, eine Ware erwirbt, zum Beispiel eine Schachtel Zigaretten, dass man sich einer Dienstleistung versichert, zum Beispiel eine Kreditkarte beschafft oder sich dem Telephonrundspruch anschliesst. In den Bereich des Kommerziellen würden aber auch all jene Dienstleistungen gehören, deren man sich im Verkehrswesen und im Tourismus bedient; man kauft nicht nur ein Air-Tikket, man kauft sich auch die Ferien am sonnigen Strand. Das «Verkaufsgespräch», welches das Plakat mit dem Betrachter führt, ist im Lauf der Zeit subtiler, man darf auch sagen: raffinierter geworden. Das Plakat, das für Sonnenöl wirbt, verkauft nicht nur einen Schutzfilter gegen Ultraviolett-Bestrahlung, es verkauft auch gesunde Bräune, darüber hinaus: Jugend, Schönheit, Anziehungskraft; letztlich verkauft es uns Ferien, Freizeit, Freiheit von Beruf und Sorgen: also Glück.

Neben dem Plakat, das kommerziell etwas anbietet, gibt es den breiten Bereich des nichtkommer-

L'affiche, messagère par excellence

Au cours de l'histoire plus que centenaire de l'affiche, diverses méthodes ont été appliquées pour présenter, dans une vue d'ensemble, toute la richesse et variété des affiches. Chacun de ces systèmes de classification – par exemple par artistes, pays d'origine, périodes de création, techniques graphiques, styles et, surtout, par secteurs d'application, sujets et motifs – donne un éclairage différent de la création d'affiches. Aucune de ces «vues» ne livre, à elle seule, des réponses satisfaisantes aux questions posées: leur combinaison, par contre, s'annonce particulièrement révélatrice.

Si l'on met l'accent sur le rôle de l'affiche, donc sur ses fonctions spécifiques dans la publicité et l'information, il en résulte tout d'abord une subdivision de l'ensemble de la production d'affiches en deux secteurs principaux: les affiches commerciales et les affiches non commerciales. Cette distinction très générale part de l'idée que le développement historique a produit des «faisceaux» de besoins difficilement comparables et que l'affiche était considérée, et est considérée aujourd'hui encore, comme un moyen adéquat pour répondre à ces besoins.

On peut considérer l'affiche comme une invitation, un message attrayant, largement diffusé grâce aux techniques de reproduction modernes et s'adressant à une masse anonyme d'individus difficiles à atteindre par d'autres moyens d'information. On peut la concevoir comme une incitation, un appel pressant à agir, même si cette action consiste simplement à satisfaire les besoins latents que l'affiche a réveillés, stimulés ou suggérés par les motifs, produits ou services qu'elle propose.

Pour l'affiche commerciale, cela signifie que l'on achète un bien de consommation, un produit, tel qu'une boîte de cigarettes, ou que l'on acquiert une prestation de service, comme par exemple une carte de crédit ou un raccordement à la télédiffusion. Le secteur commercial comprend également toutes les prestations des services de transport et du tourisme: on n'achète pas seulement un billet d'avion, mais encore des vacances sur une plage ensoleillée. La «conversation de vente» engagée par l'affiche avec l'intéressé est devenue plus subtile avec le temps, plus raffinée, pourrait-on même dire. L'affiche publicitaire qui présente une huile solaire ne vend pas seulement un filtre de protection contre les rayons ultraviolets, mais encore un bronzage parfait, signe de santé, et bien d'autres atouts encore – jeunesse, beauté, attrait irrésistible, voire même vacances, loisirs, vie libérée des contraintes professionnelles et des soucis – en un mot: le bonheur.

A côté de l'affiche qui présente une offre commerciale, il y a le vaste secteur de l'affiche non com-

gency, such warning poster-envoys were even more forceful: *"Wer nicht schweigt, schadet der Heimat"* (meaning the same as *"Careless talk costs lives"*) warned Swiss citizens against foreign agents during the Second World War.

Commercial posters for products, firms and whole branches of industry, for private and public services, can be compared with a great variety of posters on the non-commercial side which are concerned with the needs of our society, with welfare, social services, humanitarian and charitable works and with political life in its widest sense. At the extreme edge of the non-commercial poster field are those official notices which draw the attention of citizens to their rights and duties. The ancestor of this type of poster is the "proclamation" of an authoritarian government.

In between the commercial and the non-commercial there is a wide intermediate zone comprising themes such as public hygiene, the whole range of sports and also that of cultural activities and events, such as theatres, concerts, films, exhibitions, newspapers and books. Is a poster with the slogan "Read a Book" an advertisement for the book trade or an invitation to the meaningful use of leisure time? Do posters for sports events serve the pleasure of the public, the formation of leisure activities or the commercial interests concerned? Do art exhibition posters serve the formation of taste and aesthetic pleasure or the improvement of the revenues of art institutes? It is hard to perceive any clear dividing lines. Posters in these sectors have cultural, social and also commercial aspects, with varying emphasis. Separate and different categories are formed by posters which can be described as political in the widest sense, above all election and plebiscite posters. Similarly independent and anti-commercial are posters concerned with humanitarian or social activities and generally calling on us to give. The donation may be money or our blood; it may also be old materials which are to be delivered for recycling — a word which refers us to a theme of increasing importance, applying also to the poster as a conveyor of messages, invitations and warnings: the environment. The accent may be upon the threat to the environment and its destruction, thus a warning, or it may concentrate on the protection of the environment, on protective measures and a change of attitudes.

What all these various and partly contradictory themes have in common is that they are addressed to the whole population; that the individual, every individual, is addressed and asked to consider his actions within the whole framework of the public life of our society, whether a certain kind of behaviour is recommended or a warning is given

ziellen Plakats. Es fordert uns liebenswürdig auf, zum Beispiel höflich und freundlich miteinander zu sein; es empfiehlt uns, bereits mit einem Unterton der Mahnung, ja indirekten Drohung, zu unserer eigenen Sicherheit und der des Landes «schon jetzt» genügend Notvorräte anzulegen; es warnt uns vor Alkohol am Steuer und vor Taschendieben. Es gab Notzeiten, da waren derartig warnende Plakatbotschaften noch eindringlicher: «Wer nicht schweigt, schadet der Heimat» hiess im Zweiten Weltkrieg die Warnung vor Spitzeln in ausländischen Diensten.

Den kommerziellen Plakaten für Waren, Firmen und ganze Industriezweige, für private und öffentliche Dienstleistungen auf der einen Seite lassen sich auf der nichtkommerziellen anderen Seite vielerlei Plakate gegenüberstellen, deren Thematik mit Bedürfnissen unserer Gesellschaft, mit Wohlfahrt, mit sozialen Einrichtungen, mit humanitären und karitativen Werken sowie mit dem politischen Leben im weitesten Sinn zu tun haben. Ganz am äusseren Rand des nichtkommerziellen Plakates stehen jene behördlichen Informationen, die den Bürger auf Rechte und Pflichten aufmerksam machen.

Zwischen dem kommerziellen und dem nichtkommerziellen Plakat gibt es eine breite Zwischenzone, in die sowohl Themen der Volkshygiene, der weite Bereich des Sports, aber auch die volle Breite der kulturellen Aktivitäten und Angebote gehören – etwa Theater, Konzert, Film, Ausstellung, Zeitung und Buch. Sind Plakate mit dem Slogan «Lies ein Buch» Branchenwerbung des Buchhandels oder Einladung zur sinnvollen Freizeitgestaltung? Dienen Plakate für Sportveranstaltungen dem Volksvergnügen, der Freizeitgestaltung oder dem einschlägigen Kommerz? Dienen Ausstellungsplakate der Geschmacksbildung und dem ästhetischen Vergnügen oder der Verbesserung der Einnahmen von Kunstinstituten? Scharfe Trennlinien sind da wohl kaum zu ziehen. Plakate in diesen Bereichen haben, mit wechselnder Gewichtung, kulturelle, soziale wie kommerzielle Aspekte.

Kategorien eigener und anderer Art sind die Plakate, die man im weitesten Sinn als politische Plakate bezeichnen kann: vor allem Wahl- und Abstimmungsplakate. In gleicher Weise eigenständig und antikommerziell sind die Plakate, die mit sozialen oder humanitären Werken zu tun haben und uns in der Regel zu Spenden auffordern. Diese Spende kann Geld sein oder unser Blut; es kann auch um Altstoffe gehen, die einem Recycling zugeführt werden sollen. Dieses Stichwort verweist auf eine Thematik, die auch im Bereich des Plakats – als eines Übermittlers von Botschaften, Aufforderungen und Warnungen – zunehmend an

merciale. C'est, par exemple, l'affiche qui nous invite gentiment à être polis et aimables dans nos rapports mutuels; elle nous recommande, sur un ton déjà plus impératif et avec une pointe de menace, de constituer «dès à présent» des réserves suffisantes pour les temps de crise; elle nous avertit des dangers de l'alcool au volant et recommande de prendre garde aux pickpockets. Il y a eu des temps de crise où ces messages d'avertissement étaient plus incisifs encore: «Qui ne sait se taire, nuit à son pays» proclamait pendant la Seconde Guerre mondiale une affiche de mise en garde contre les espions au service des puissances étrangères.

Si les affiches commerciales se concentrent sur des produits, des entreprises et des secteurs économiques complets, ainsi que sur les services privés et publics, les affiches à caractère non commercial s'étendent à une vaste gamme d'activités, à des sujets liés aux besoins de notre société, à la prévoyance et aux institutions sociales, aux œuvres humanitaires et caritatives, ainsi qu'à la vie politique au sens le plus large du terme. A la limite extrême de l'affiche non commerciale se situent les informations des pouvoirs publics visant à rappeler aux citoyens leurs droits et leurs devoirs. Ses ancêtres étaient les avis et décrêts «officiels» de jadis.

Entre les affiches commerciales et non commerciales, il existe une vaste zone intermédiaire, qui englobe tant les sujets relatifs à l'hygiène publique et au vaste domaine des sports que toute la gamme des activités et manifestations culturelles, telles que théâtre, concert, cinéma, expositions, livres et journaux. Est-ce qu'une affiche avec le slogan «Lisez un livre» constitue une publicité spécifique au secteur de la librairie ou bien un appel pour un aménagement judicieux des loisirs? Est-ce qu'une affiche annonçant une manifestation sportive est placée au service du divertissement populaire, de l'aménagement des loisirs ou du commerce local? Est-ce que les affiches d'exposition contribuent à former le goût du public et à la jouissance esthétique ou à améliorer les recettes des institutions exposant des œuvres d'art? Il est difficile de tracer des lignes de démarcation précises. Les affiches de cette zone intermédiaire présentent des aspects tant culturels que sociaux, voire même commerciaux, avec une pondération variable.

Une catégorie spécifique et bien distincte d'affiches est constituée par celles à caractère politique, au sens le plus large du terme: il s'agit d'affiches électorales et de propagande, diffusées au moment des votes. Le caractère anticommercial et autonome est également typique pour les affiches

against the consequences of its opposite. It is not always clear what is meant and the actual themes are often ambiguous. *"Brauch Gas"* ("Use Gas") — does this support international agreements on natural gas, is it meant to increase the sales of a comparatively "clean" form of energy or reduce the need for atomic power?

One can well imagine that the "environment" theme is becoming increasingly important at a time when the endangering of our environment and therefore of our basic conditions of life takes on ever more menacing forms. It is reasonable to suppose that more and more efforts would be made to influence people in general to behave sensibly, using the medium of the poster. Examples of this policy are increasing. A growth of this kind in non-commercial posters might even go together with a reduction in commercial posters. The prohibitions of alcohol and tobacco advertising which have come into force in certain areas point in this direction. It is also — theoretically — possible that a (hypothetical) poster campaign for weekend residences in unspoilt country might be prohibited, if only to avoid arousing the anger of the environmentalists.

In the fascinating history of the Swiss poster, important artistic achievements are to be found in all branches. There is no field of poster work which can be judged "unworthy" because of its theme, its place in a special commercial or social sector or the special character of its public. A convincing and even an outstanding design is possible in any field. The best Swiss poster designers have repeatedly proved this point in the course of their careers, which have not excluded "lesser" commissions.

Nevertheless, the question remains open whether the practical pressures of commercial poster work and the many and complex conditions associated with modern advertising campaigns, compared with the much greater freedom of non-commercial work, do not make the latter more stimulating for an original artistic design. There is no clear affirmative or negative answer to the question: Are poster masterpieces the result of fully used freedom or of the triumphant overcoming of external pressures? The question might be indirectly elucidated by the easily verified fact that the high points of the Swiss poster are to be found precisely in those decades of the Fifties and Sixties when extremely impressive posters were created even for the most banal and insignificant things. That the poster for a masterpiece of cinema art or the Shakespeare season of a municipal theatre, and also the one for high-quality silks, should be of above-average formal quality — that can almost be taken for granted.

Bedeutung gewinnt: die Umweltthematik. Gemeinsam ist all diesen verschiedenartigen, zum Teil widersprüchlichen Themen, dass sie die Gesamtgesellschaft angehen; dass der Einzelne, jeder Einzelne angesprochen ist, sein Verhalten, sein Tun im Gesamtgeschehen des öffentlichen Lebens unserer Gemeinschaft zu sehen, ob nun ein bestimmtes Verhalten empfohlen wird oder ob vor dessen Konsequenzen gewarnt wird. Nicht immer ist klar, was gemeint ist, oft bleiben die eigentlichen Motive im dunkeln. «Brauch Gas» – soll das zur Sicherung oder Finanzierung internationaler Erdgas-Vereinbarungen beitragen, soll es einer vergleichsweise «sauberen» Energie vermehrten Absatz bringen oder die Notwendigkeit von Atomenergie reduzieren?

Man könnte sich vorstellen, dass der Aufgabenbereich «Umwelt» beim Plakat zunehmend an Bedeutung gewinnt, dies in einer Zeit, in der die Gefährdung unserer Umwelt und damit unserer elementaren Lebensbedingungen immer bedrohlichere Formen annimmt. Es wäre denkbar, dass durch das Mittel des Plakats vermehrt versucht wird, die Allgemeinheit zu vernünftigem Verhalten anzuregen. Beispiele dafür mehren sich. Eine solche Zunahme der nichtkommerziellen Plakate könnte sogar mit einer Reduktion der kommerziellen einhergehen. Die sich mehrenden Verbote der Alkohol- und Tabakwerbung weisen in diese Richtung.

In der spannenden Geschichte des Schweizer Plakats finden sich in allen Gattungen bedeutende künstlerische Leistungen. Es gibt keinen Plakatbereich, der durch seine Thematik, durch seine Zugehörigkeit zu einem speziellen Wirtschafts- oder Sozialbereich, durch die Eigenart seines Zielpublikums von vornherein «unwürdig» und einer formal überzeugenden, ja gar überragenden Gestaltung nicht zugänglich wäre. Die grossen Plakatgestalter der Schweiz haben dies im Lauf ihrer Karriere, die auch «niedere» Aufgaben nicht ausgespart hat, wiederholt unter Beweis gestellt.

Allerdings bleibt die Frage offen, ob beim kommerziellen Plakat die Sachzwänge und die vielen – im Rahmen heutiger Werbestrategien besonders komplexen – Bedingungen oder aber die vergleichsweise viel grösseren Freiheiten beim nichtkommerziellen Plakat stimulierender sind für eine originale künstlerische Gestaltung. Es gibt kein deutliches Ja oder Nein auf die Frage: Sind Meisterplakate Resultat voll genutzter Freiheit oder siegreicher Überwindung auferlegter Zwänge? Die Frage liesse sich vielleicht indirekt mit der (durch überprüfbare Fakten leicht zu belegenden) Tatsache klären, dass die Höhepunkte des Schweizer Plakats gerade in jenen fünfziger und sechziger Jahren liegen, da selbst für die banalsten, belang-

ayant trait à des œuvres sociales ou humanitaires: l'objectif de ces affiches est de recueillir des dons, qu'elles s'adressent à des donateurs de fonds ou à des donneurs de sang; il peut même s'agir simplement de recueillir des matières usées pouvant être recyclées. Le terme de recyclage évoque un domaine dans lequel l'affiche – comme messagère chargée de transmettre des informations, appels et avertissements – ne cesse de gagner en importance: le domaine de l'écologie. L'accent peut être mis sur la dégradation et la destruction de l'environnement, donc se traduire par un cri d'alarme, ou sur la nécessité de protéger l'environnement, donc de préconiser les mesures qui s'imposent ou un changement d'attitude.

Le trait commun de tous ces sujets, souvent contradictoires, est qu'ils intéressent l'ensemble de la société, que chaque membre de cette société doit se sentir personnellement concerné, contraint de revoir sa manière d'être et d'agir à la lumière de la vie publique de notre communauté. L'affiche peut exprimer tantôt la recommandation d'adopter un certain comportement, tantôt une mise en garde contre les conséquences néfastes d'une attitude erronée. L'intention véritable n'est pas toujours clairement exprimée, souvent les vrais motifs restent diffus. «Le gaz à votre service»: s'agit-il de garantir ou de financer les conventions internationales relatives au gaz naturel, d'augmenter la consommation d'une énergie qualifiée de «propre» ou de la nécessité de réduire la part de l'énergie nucléaire?

On peut s'imaginer que l'affiche consacrée au domaine de l'écologie continue de gagner en importance, en particulier à une époque où la menace qui affecte notre environnement et, partant, nos conditions élémentaires de vie, prend des formes toujours plus alarmantes. Il est concevable que l'affiche serve encore davantage de moyen pour inciter tous et chacun à adopter une attitude raisonnable. Les exemples de ce genre se multiplient. Une telle augmentation des affiches non commerciales pourrait même aller de pair avec une réduction des affiches commerciales. Les interdictions de toute publicité pour l'alcool et le tabac pointent dans cette direction. L'on pourrait – théoriquement – s'imaginer aussi que l'on renonce à toute publicité (hypothétique) en faveur des résidences secondaires dans les régions de montagne encore peu touchées, simplement pour ne pas susciter la colère des écologistes par de telles affiches.

L'histoire passionnante de l'affiche suisse est riche en productions artistiques dans tous les genres qui existent. Aucun secteur de la création d'affiches n'est de prime abord «indigne» ou «incapable»

But the fact that a poster artist can make washing powder or mineral water the occasion for a masterpiece, calls for further consideration.
Independent graphic designers, especially free artists, have always preferred to take on jobs with the greatest possible scope for development, ever since the beginnings of the artistic poster around 1900. These are posters in which the pressures of advertising economics are not decisive, where originality in ideas and form is possible and perhaps even expressly requested by the client.
Swiss poster history shows that the designers of bold non-commercial posters have again and again had the opportunity, thanks to the prestige gained thereby, to take great liberties in commercial poster work, and thus achieved remarkable results in that field as well. When this experience is repeated with a certain regularity, it becomes possible to describe the non-commercial poster as a trend-setter in design or as an aesthetic "leader".
The non-commercial posters here referred to may be described as "social" and "political". What they have in common is that the clients' aims are, in the final analysis, idealistic. These clients include, on the one hand, institutions, associations, foundations and working groups of a charitable, social, medical, religious or similar character; and on the other hand, councils, ministries, political parties and lobbies with various aims, or *ad hoc* pressure groups formed by associations of residents and citizens for a specific purpose.
These posters publicize concerns which are of great importance to the sponsors, precisely determined and sometimes highly specialized. They are addressed to all and sundry. An appeal is made to the vigilance and curiosity of the individual, to his consideration of the content of the message. The reaction to the poster appeal depends on the personal character of the viewer, his own interests, likes and dislikes, his opinions and knowledge, his social status and his general frame of mind.
If interest is awakened and consideration aroused through the special theme of the poster, it has the potential to influence decisions through its verbal and pictorial expression. The use of market research may be able to explain which poster messages have a stimulating effect on the activities of the viewer, which are inhibiting or even counterproductive, which pictorial envoys are found attractive and which are negative in their effect. Such questionnaires must also take the surroundings of the poster into account in order to avoid false conclusions.
Those who are addressed by the poster may disregard its message, either instinctively or deliberately. They can resist its appeal by deciding "This

losesten Dinge grossartig einprägsame Plakate geschaffen wurden. Dass das Plakat für ein Meisterwerk der Filmkunst oder den Shakespeare-Zyklus eines städtischen Theaters, aber auch für hochmodische Seidenstoffe seinerseits von überdurchschnittlicher formaler Qualität sei – das darf man beinahe als Selbstverständlichkeit voraussetzen. Dass ein Plakatgraphiker aber Waschpulver oder Mineralwasser zum Anlass für ein Meisterplakat macht, verdient Beachtung.
Eigenständige graphische Gestalter, vor allem auch freie Maler, haben sich seit den Anfängen des künstlerischen Plakats um 1900 mit Vorliebe den Aufgaben mit möglichst grossem Entfaltungsspielraum zugewandt: Plakaten also, bei denen werbewirtschaftliche Zwänge nicht ausschlaggebend sind, Plakaten, bei denen ideenmässig und gestalterisch-formal neuartige Lösungen möglich, ja vom Auftraggeber ausdrücklich gefragt sind.
Die Schweizer Plakatgeschichte zeigt, dass immer wieder Entwerfer kühner nichtkommerzieller Plakate dank dem mit solchen Arbeiten erworbenen Prestige sich auch im Bereich des kommerziellen Plakats grosse Freiheiten erlauben durften und damit auch hier ungewöhnliche Leistungen vollbracht haben. Wenn sich dies mit einer gewissen Regelmässigkeit wiederholt, dann darf das nichtkommerzielle Plakat im Gestalterischen als eine Art «Trendsetter» oder ästhetische «Leitfigur» bezeichnet werden.
Die hier anvisierten nichtkommerziellen Plakate sollen als «soziale» und als «politische» Plakate bezeichnet werden. Gemeinsam ist ihnen, dass die Auftraggeber letztlich ideelle Ziele anstreben.
Diese Auftraggeber sind einerseits Institutionen, Vereine, Stiftungen, Arbeitsgemeinschaften karitativen, sozialfürsorgerischen, präventivmedizinischen, kirchlichen oder ähnlichen Charakters, anderseits aber Behörden, Amtsstellen, politische Parteien oder politische Arbeitsgruppen, Interessengemeinschaften unterschiedlicher Zielsetzung oder ad hoc zusammengetretene «Betroffenengruppen», wie man sie von Deutschland her als «Bürgerinitiativen», bei uns eher als «Anwohner- oder Anstössergemeinschaften» kennt.
Diese Plakate verkünden jeweils ein den Auftraggebern zentral wichtiges, genau determiniertes, manchmal extrem spezielles Anliegen. Angesprochen ist jeder und jede. Es wird an die Wachheit und Neugier des Individuums appelliert, an sein Nachdenken über den Inhalt der Botschaft. Von der Persönlichkeitsstruktur des Angesprochenen, seinen Interessen, Neigungen und Abneigungen sowie seiner sozialen Lage hängt die Reaktion auf den Plakataufruf ab.
Ist das Interesse geweckt, das Nachdenken durch

d'aboutir à une conception formelle convaincante, voire même exceptionnelle, en raison soit des sujets traités, soit de son appartenance à un secteur économique ou social spécifique, ou encore des particularités de son public cible. Les grands affichistes suisses l'ont prouvé tout au long de leur carrière: jamais ils n'ont exclus de leur activité créatrice des sujets moins «nobles».
Il se pose toutefois la question de savoir laquelle des deux catégories d'affiches paraît plus propice à la création d'œuvres artistiques originales: est-ce l'affiche commerciale, avec ses différentes contraintes et les multiples conditions à remplir en raison des stratégies publicitaires toujours plus complexes, ou bien l'affiche non commerciale, avec ses libertés beaucoup plus larges? Il n'y a pas de réponse définitive à la question ainsi posée: les affiches hors pair sont-elles le résultat d'un épanouissement total en pleine liberté ou l'aboutissement des contraintes surmontées avec succès?
La question pourrait peut-être trouver une réponse indirecte par le fait (facile à établir grâce à des preuves confirmées) que les moments d'apogée de l'affiche suisse se situent précisément dans les années cinquante et soixante, où des affiches exceptionnellement suggestives ont été créées, même pour les choses les plus banales. Qu'une affiche pour un chef-d'œuvre de l'art cinématographique ou un cycle de représentations d'œuvres de Shakespeare par un théâtre municipal, ou encore pour des soieries en vogue, soit nécessairement de qualité formelle exceptionnelle – voilà qui paraît l'évidence même. Qu'un graphiste réalise un chef-d'œuvre en créant une affiche pour une poudre à lessive ou une eau minérale mérite d'être spécialement relevé.
Dès les débuts de l'affiche artistique en 1900, les créateurs graphiques indépendants, et plus spécialement les peintres, se sont concentrés de préférence sur les réalisations offrant un maximum de liberté d'épanouissement, c'est-à-dire sur les affiches sans déterminantes publicitaires contraignantes, permettant de développer de nouvelles idées et des conceptions formelles originales, parfois même sur demande explicite du commettant.
L'histoire de l'affiche suisse révèle que les créateurs de solutions audacieuses pouvaient, grâce au prestige acquis par des travaux non commerciaux, bénéficier de larges libertés aussi dans le domaine de l'affiche commerciale, où ils ont créé des œuvres exceptionnelles. Si ce phénomène se produit avec une certaine régularité, l'affiche non commerciale peut être considérée comme «trend setter» dans la conception formelle: c'est elle qui donne le ton sur le plan esthétique.
Les affiches non commerciales visées ici peuvent

is not for me". On the other hand, those who react positively to the poster are hardly likely to send off an immediate payment or make a firm voting decision in obedience to the poster's recommendation. Nevertheless, in ideal circumstances, the poster message — in association with other channels of information and other media, and over a period of time — has the power to make such an impression concerning the matter in hand, that it is almost bound to have an influence on people's actions.
In the field of social, humanitarian or preventive actions, the poster is generally only one of several possible methods of making people aware of a regional or national concern. However, when supported by a previous or current publicity campaign, it is able to repeat and hammer home the essential message in a condensed form, particularly through the thematic and also the formal power of its pictorial elements.
The political poster commonly uses less differentiated verbal and pictorial methods. The words are easily understood, the picture presented with utmost simplicity and vivid directness. An appeal is made to subliminal feelings and emotions are released, sometimes of an aggressive character. Elements of caricature and pictorial and verbal wit are also used. The voter is courted, led on in a subtle way and an appeal is even made to his weaknesses; or else he is imperatively addressed, directly and point-blank.
In plebiscites (which are common in Switzerland) on specific matters such as women's voting rights, working hours, abortion law reform and so on, the posters reflect the intensity with which such questions are publicly discussed and disputed. The basis of objective reality is often abandoned — also in the poster — and exaggerations, even defamations, lead the field here and there in words as in pictures. The shamelessness of the political poster may well be one reason why, contrary to normal practice in other branches of poster art, the designer often chooses to remain anonymous.
It is the political poster, in fact, that is a barometer of the state of a nation, of the virulence of its political life, of its freedom of expression and the fairness or recklessness of its political conflict. Any Swiss who examines the poster landscape before an important election or plebiscite should feel glad of the liveliness, the intensity and the expenditure with which preparations are made for decisions by the people. The lack of mockery and ironical wit (often the deadliest weapon) in a political struggle which is often all too solemn, is rather a reflection of the country's political style than of insufficient imagination in poster art.
A question often discussed is whether, and how

die spezielle Thematik in Gang gebracht, so vermag das Plakat durch die verbale und bildliche Artikulation seiner Botschaft Entscheidungen zu beeinflussen. Vielleicht vermöchten Umfragen zu klären, welche Plakataussagen auf das Tätigwerden der Angesprochenen stimulierend, welche hemmend oder gar verhindernd gewirkt haben, welche Bildbotschaft als attraktiv, welche als abstossend empfunden wurde. Bei solchen Meinungsumfragen müsste allerdings auch das Umfeld mitberücksichtigt werden, um Kurzschlüsse auszuschalten.
Der durch das Plakat Angesprochene kann dessen Aufforderung instinktiv oder willentlich missachten. Er kann mit einem «Ich-nicht» Widerstand leisten. Wer positiv auf das Plakat reagiert, wird allerdings kaum sofort seinen Einzahlungsschein zur Post bringen oder im Sinne der Plakatempfehlung seine Wahl- oder Abstimmungspapiere ausfüllen. Die Aufforderung des Plakats vermag jedoch im Idealfall das – zusätzlich durch andere Medien und in zeitlicher Staffelung – vorgebrachte Anliegen so mitzuprägen, dass fast zwanghaft im Sinne der Plakatbotschaft gehandelt wird.
Im Bereich der sozialen, humanitären oder präventiven Aktionen ist das Plakat in der Regel nur eine von mehreren genutzten Möglichkeiten, die Menschen für ein regionales oder landesweites Anliegen zu sensibilisieren. Es vermag jedoch, auf einen bereits erfolgten oder in Gang befindlichen Feldzug abgestützt, schlagwortartig kondensiert das Wesentliche zu repetieren oder einzuhämmern, dies vor allem durch die motivische und meist auch formale Kraft seiner Bildelemente.
Das politische Plakat benutzt häufig weniger differenzierte Sprach- und Bildmittel. Die Sprache ist allgemeinverständlich, lapidar, das Bild auf bildbogenhafte Einfachheit und anschauliche Direktheit angelegt. Es wird an unterschwellige Regungen appelliert, und es werden Emotionen, manchmal Aggressionen ausgelöst. Auch karikaturistische Elemente, der Bildwitz und der Wortwitz werden eingesetzt. Der Wähler wird umworben, raffiniert gegängelt und auch in seinen Schwächen getroffen, oder aber er wird sehr direkt, ohne Umschweife, imperativ angeredet.
Bei Abstimmungen über Sachfragen, wie Frauenstimmrecht, Arbeitszeit, Schwangerschaftsabbruch usf., spiegeln die Plakate die Intensität, mit der solche Fragen in der Öffentlichkeit diskutiert oder umkämpft werden. Der Boden der Sachlichkeit wird – auch im Plakat – oft verlassen, Übertreibungen, sogar Verunglimpfungen im Wort wie im Bild beherrschen hin und wieder das Feld. Die Unzimperlichkeit des politischen Plakats mag mit ein Grund dafür sein, dass es den Entwerfer oft die An-

être qualifiées d'affiches «sociales» et «politiques». Leurs commettants poursuivent tous des buts idéologiques. Ce sont, d'une part, des institutions, associations, fondations, communautés de travail d'œuvres caritatives, de la prévoyance sociale, de la médecine préventive, des milieux ecclésiastiques et d'autres organisations semblables et, d'autre part, des autorités, services publics, partis ou groupes de travail politiques, communautés d'intérêt à buts divers ou «groupements de personnes concernées», constitués ad hoc, tels que les «initiatives de citoyens» en Allemagne ou les «communautés de riverains ou de bordiers» en Suisse.
Ces affiches propagent un message bien déterminé, d'importance capitale pour le commettant, et de caractère parfois très spécifique. L'affiche s'adresse à chacun, homme ou femme, accroche l'attention, éveille la curiosité, incite à réfléchir au contenu du message. La personnalité du destinataire, ses intérêts personnels, ses préférences, ses aversions, ses conceptions et ses connaissances antérieures, sa situation sociale et son état psychique déterminent sa réaction à l'appel lancé par l'affiche.
Dans le domaine des actions sociales, humanitaires ou préventives, l'affiche ne constitue en règle générale qu'une possibilité parmi d'autres pour susciter l'intérêt pour des sujets à caractère régional ou national. Elle offre toutefois l'avantage de répéter de manière percutante et concise les points essentiels d'une campagne déjà terminée ou encore en cours, en se servant de slogans incisifs et, surtout, d'éléments formels suggestifs, empreints d'une grande force d'expression visuelle.
L'affiche politique se sert souvent d'une langue et d'images moins différenciées. Le texte est facile à comprendre, lapidaire, l'image est illustrative, de portée directe. L'appel vise à réveiller des sentiments cachés, à déclencher des émotions, parfois même des agressions. Des éléments caricaturaux, des dessins et textes humoristiques entrent en jeu. L'électeur est entouré, envoûté, sollicité jusque dans ses points faibles, ou bien abordé très directement, sans ambages, sur un ton impératif.
Lors de votes sur des questions spécifiques, telles que le droit de vote des femmes, la durée du travail, l'interruption de la grossesse, les affiches reflètent l'intensité des discussions et contestations engendrées par de telles questions au sein du public. L'objectivité est abandonnée – aussi pour l'affiche – au profit d'exagérations démesurées, voire même de dénigrements par le texte et par l'image. Le fréquent manque de subtilité de l'affiche politique explique peut-être pourquoi, à l'en-

far, the practical function of a poster determines its style. Put in more general terms, the question is, how far the character or climate of a given field of activity — be it sport, industrial exhibitions or fashion, charity appeals, soft drinks or washing powder — shows itself in the design of the poster, independently of the personal style of the designer and of the style currently in fashion.

Ernst Keller, who as founder of the first specialist graphics class at a Swiss school of arts and crafts has trained several generations of graphic artists in Zurich, once said: "A poster cannot be really good unless its theme is effectively presented not only in the choice of subject and in the text message, but also in the style of colour and form. In the ideal case it must be possible to grasp the poster's message intuitively and spontaneously, without any mental effort by the viewer. Thus a poster advertising coffee must create the idea of coffee simply through its colours or colour combinations and the evocative formal character of its picture and text elements. By association, the viewer must 'taste the coffee' in such a poster."

That is certainly correct, but the poster message cannot always be reduced to such a common denominator in all the jobs that a poster designer has to undertake. In a competitive climate of great complexity, even comparatively simple jobs are today dependent upon difficult secondary considerations. In addition, the poster designer's personal characteristics, his professional conditions and his artistic concepts have also been transformed in a rapidly changing world.

If Ernst Keller's view were still correct, it would follow that in similar or even identical commissions, the designs of various artists would approximate very closely to one another. As the selection of posters in this publication shows, that is very far from being the case. However, Ernst Keller was probably also aware that not only does every blend of coffee taste different, but each roasting produces different nuances of aroma, and finally — certainly the most important point — everyone prepares coffee in his own individual way.

One of the fascinating aspects of poster art is that it provides a great variety of information about the country in which, and for which, it is created. The poster speaks the language of the people to whom it is addressed. To put it in simple terms: the more a poster is tailor-made for its mission, the more successful it will be. It is therefore often difficult to judge the poster production of a foreign country correctly, that is to say, when one does not have sufficient precise knowledge of the situation to which the poster refers, whether directly or subliminally, as through a veil.

onymität wählen lässt, die unpersönliche Botschaft. Gerade das politische Plakat ist ein Gradmesser für den Zustand einer Nation, ein Gradmesser für die Virulenz des politischen Lebens, für die freie Meinungsäusserung und die Fairness oder Rücksichtslosigkeit in der politischen Auseinandersetzung. Wer in der Schweiz vor wichtigen Wahlen oder Abstimmungen die Plakatlandschaft durchwandert, darf sich über die Lebendigkeit, die Intensität, den Aufwand freuen, womit Volksentscheide vorbereitet werden. Dass der liebenswürdige Spott, die geistvolle Ironie (oft als tödlichste Waffe) in einem häufig allzu «bierernsten» politischen Kampf auf den Plakatwänden fehlen, ist eher ein Spiegel des politischen Stils als mangelnder plakatkünstlerischer Flexibilität in diesem Land.

Es wird oft die Frage erörtert, ob und wieweit die konkrete Aufgabe den Stil eines Plakats bestimme. Oder, genereller gefragt, wieweit der Charakter, das «Klima» eines bestimmten Aufgabenkreises – zum Beispiel Sport, Messe oder Mode, Sammelaktion eines Hilfswerks, Erfrischungsgetränk oder Waschmittel – die Plakatgestaltung mitpräge, unabhängig vom persönlichen Stil des Gestalters, unabhängig sogar vom jeweiligen «Zeitstil».

Ernst Keller, der als Begründer der ersten Fachklasse für Graphiker an einer Schweizer Kunstgewerbeschule mehrere Generationen von Graphikern in Zürich ausgebildet hat, sagte: «Ein Plakat ist erst dann wirklich gut, wenn seine Thematik nicht nur in der Motivwahl, nicht nur in der Textbotschaft wirkungsvoll präsentiert ist, sondern auch im farbformalen Stil sich bekundet.» Im Idealfall müsse die Botschaft des Plakats, ohne gedanklichen Nachvollzug durch den Betrachter, spontan und intuitiv erfasst werden können. Ein Kaffeeplakat müsse also allein schon durch die Farben oder Farbkombinationen, durch suggestive Formcharaktere der Bild- und Textelemente das Klima von Kaffee erzeugen. Assoziativ müsse der Betrachter in einem solchen Plakat den «Kaffee schmecken». Das ist gewiss richtig. Nicht bei allen Aufgaben, die sich dem Plakatgraphiker stellen, lässt sich allerdings die Botschaft auf einen solchen elementaren Nenner reduzieren. In einer sehr komplexen Wettbewerbssituation ist heute selbst die vergleichsweise einfache Aufgabe von schwierigen Randbedingungen abhängig. Anderseits haben sich auch die Persönlichkeitsstrukturen, die fachlichen Voraussetzungen wie die gestalterischen Konzepte der Plakatgraphiker in einer rasch sich verändernden Welt gewandelt.

Hätte Ernst Keller noch immer recht, so müssten bei gleichartiger oder gar gleicher Aufgabenstellung die Entwürfe verschiedener Gestalter sich sehr nahe kommen. Das ist – wie gerade die Pla-

contre des créateurs d'autres catégories d'affiches, l'auteur de l'affiche politique se réfugie souvent dans l'anonymat.

Or, l'affiche politique est précisément le reflet de l'état d'une nation, l'expression de la virulence de sa vie politique, de sa liberté d'expression, de l'esprit de fairness ou de désinvolture dans la confrontation politique. Celui qui passe devant les murs garnis d'affiches avant d'importantes «votations» ou élections en Suisse, se réjouira de constater la vivacité, l'intensité, l'ardeur avec laquelle on prépare les décisions du «souverain». L'absence de satire enjouée, d'ironie spirituelle (souvent l'arme la plus meurtrière) sur les murs d'affichage pendant un lutte politique «sérieuse jusqu'à l'ennui», traduit plutôt le style politique de ce pays qu'un manque de flexibilité dans la création d'affiches politiques. Est-ce que le rôle concret assigné à une affiche est déterminant pour le style? Cette question, souvent posée, consiste à demander dans quelle mesure le caractère, le «climat» d'un certain secteur d'activité – par exemple sports, exposition ou mode, collecte de fonds, boisson rafraîchissante ou produit de lessive – marque la conception formelle de l'affiche, indépendamment du style personnel de l'affichiste, voire même du style «en vogue» à ce moment précis.

Ernst Keller, fondateur de la première classe professionnelle de graphistes à une école des arts et métiers suisse et qui a formé plusieurs générations de graphistes à Zurich, affirme: «Une affiche n'est véritablement bonne que si la présentation du sujet est percutante, non seulement par le choix du motif ou le message du texte, mais encore par le style exprimé à travers les formes et couleurs choisies. Dans le cas idéal, le message de l'affiche doit pouvoir être saisi spontanément, intuitivement, sans effort de réflexion supplémentaire. Une affiche proposant du café doit donc suggérer l'ambiance du café par le choix des couleurs et combinaisons de couleurs, par des éléments formels, des images et textes suggestifs incitant, par simple association d'idées, à «sentir l'odeur du café».»

Cette affirmation est certainement correcte. Mais compte tenu des nombreuses tâches que doit affronter le graphiste, il n'est pas toujours possible de ramener le message de l'affiche à un dénominateur aussi simple. Même les tâches les plus élémentaires sont aujourd'hui assorties de conditions marginales difficiles, en raison d'une situation concurrentielle toujours plus complexe. Il est vrai que dans ce monde en mutation rapide, le graphiste apporte lui aussi d'autres qualités personnelles, aptitudes professionnelles et conceptions formelles pour la création d'affiches.

Si l'assertion d'Ernst Keller valait toujours, il fau-

As an example: a poster by Rolf Gfeller, advertising the Swiss national lottery in 1966, shows shirt-sleeved players of the game of "Hornuss" in an expressively picturesque way; they are throwing their black Hornuss boards up in the air in order to intercept the red four-leafed clover, which is the good-luck symbol of a winner in the lottery. For those with no knowledge of the ancient Bernese team-game called "Hornuss" (a play on the German word for "hornet"), the meaning of the poster remains obscure and the attraction of its analogy is lost. Only those who know and have witnessed this highly traditional Swiss popular sport (in which the "strikers" team drives the Hornuss "puck" from an iron rest with a flexible club and the "stoppers" team tries to intercept the puck by throwing Hornuss "boards" in the air) can fully understand the poster's meaning: a non-verbal invitation to buy lottery tickets.

Even in one's own country, much that is often very important, even surprising and unknown, concerning the familiar world, which one thought one knew, can be read and deduced from the poster. This ability to act as a witness of the living conditions and character of a people belongs, on the one hand, to the conventional and not very articulate average poster, which, as it were, echoes its environment in order to ingratiate itself with it. On the other hand, traces of local peculiarity and the signs of regional character are also to be found in the unconventionally designed work of independent poster artists. In this way the poster, over and above its actual function, harbours an analysis, an interpretation, perhaps also a critical illumination of the world of its origin and its social environment. Occasionally its very effectiveness is due to the return into consciousness of facts or feelings from a half-forgotten, recent or more distant past. The insights into national or local mentality are all the more precise and substantial, the more the possibility exists of viewing a number of posters at the same time, preferably with a common or similar theme. Here are two extremely simplified examples:

If the posters of a given country include practically no advertisements for products, or none at all, it is clear that the lack of commercial advertising indicates the lack of a free market economy. Where "planning" posters dominate, the production and distribution of goods are not determined by individual or stimulated demand.

A lack of film posters might indicate that, in the country concerned, there is no appreciable interest in the cinema (an absurd conclusion); or, on the contrary, that cinema-going is so widespread, so much taken for granted, that poster advertising is

katauswahl in dieser Publikation zeigt – in keiner Weise der Fall. Aber wahrscheinlich wusste auch Ernst Keller, dass nicht nur jede Kaffeemischung anders schmeckt, sondern jede Röstung andere Nuancen des Aromas ergibt und dass schliesslich jeder seinen Kaffee auf individuell verschiedene Weise zubereitet.

Es gehört zu den faszinierenden Eigenschaften der Plakate, dass sie vielfältige Auskunft über das Land geben, in dem sie und für das sie hergestellt worden sind. Das Plakat spricht die Sprache der Menschen, an die es sich wenden will. Überspitzt gesagt: Je mehr das Plakat nach Mass geschneidert ist, um so erfolgreicher erfüllt es seine Mission. Oft hält es daher schwer, als Fremder die Plakatproduktion eines anderen Landes richtig zu beurteilen; dann nämlich, wenn nur ungenügend genaue Kenntnis der Gegebenheiten vorhanden ist, auf die das Plakat direkt oder unterschwellig, also verhüllt Bezug nimmt.

Ein Beispiel: Aus dem Jahre 1966 gibt es ein Plakat der Schweizer Landeslotterie von Rolf Gfeller, das in expressiver malerischer Manier hemdsärmelige «Hornusser» zeigt; sie werfen ihre schwarzen Bretter auf, um das rote vierblättrige Kleeblatt (Glückssymbol der Landeslotterie für einen Treffer) aus der Luft herunterzuholen. Ohne Kenntnis des altbernisch-bäuerlichen Mannschaftsspiels des Hornussens (eine Anspielung auf die gefährliche, grosse Wespenart der Hornisse) bleibt der Sinn des Plakats dunkel, der Reiz seiner Analogie geht verloren. Nur wer diesen traditionsreichen Schweizer Volkssport kennt und einmal erlebt hat, wie die Partei der «Schläger» den Hornuss mit dem elastischen Stecken vom eisernen Bock wegschlägt und wie die Partei der «Abtuer» mit den hochgeworfenen Hornussschindeln das puckartige Geschoss abzufangen versucht, versteht voll den Sinn des Plakats: eine nichtverbale Einladung zum Loskauf.

Selbst im eigenen Land ist aus den Plakaten vieles, oft sehr Wesentliches, manchmal auch Überraschendes, Unbekanntes herauszulesen und herauszuhören über die heimatliche Welt, die man zu kennen glaubt. Diese Fähigkeit zur Zeugenschaft für die Lebensbedingungen und den Charakter eines Volkes ist einerseits dem konventionellen, wenig artikulierten Durchschnittsplakat eigen, das sozusagen seiner Umgebung nachplappert, um sich bei ihr einzuschmeicheln. Anderseits finden sich die Spuren des Lokalen, die Merkmale regionaler Eigenart auch bei unkonventionell gestalteten Arbeiten eigenständiger Entwerfer. Da birgt das Plakat – ausserhalb der eigentlichen Aufgabe – eine Analyse, eine Deutung, vielleicht auch eine kritische Beleuchtung von Herkunftswelt und sozialer Umgebung. Gelegentlich macht es geradezu seine

drait qu'en cas de fonctions semblables ou même identiques assignées à l'affiche, les projets de divers graphistes se rapprochent très fortement. Or, comme le montrent les affiches publiées dans le présent ouvrage, une telle ressemblance n'existe nullement. Et puis, Ernst Keller savait certainement que non seulement chaque mélange de café est différent, mais aussi que chaque torréfaction produit un autre arôme et, enfin – ce qui est sans doute déterminant – que chacun a sa manière personnelle de préparer son café.

Une des propriétés les plus fascinantes de l'affiche est qu'elle fournit de multiples renseignements, tant sur le pays où elle est produite que sur le pays de destination. L'affiche parle la langue des hommes auxquels elle s'adresse. On pourrait même aller jusqu'à dire: plus une affiche est faite sur mesure, et mieux elle remplit sa mission. Il est de ce fait souvent difficile, en tant qu'étranger, de juger correctement des affiches produites par un autre pays, surtout sans connaissances précises des données auxquelles l'affiche se réfère, directement ou par le biais de sous-entendus.

L'affiche pour la Loterie nationale suisse, réalisée par Rolf Gfeller en 1966, illustre bien cette constatation. Dans un style pictural expressif, elle montre des joueurs de «Hornussen» aux bras noueux, qui lancent en l'air leurs planches noires pour décrocher dans le ciel le trèfle rouge à quatre feuilles (symbole de chance de la loterie nationale pour un numéro gagnant). Seule la connaissance du «Hornussen» (jeu d'équipe correspondant à une ancienne tradition paysanne bernoise et inspirée par le vol des «frelons») permet de comprendre le sens profond de l'affiche et de saisir le charme de l'analogie. Il faut avoir assisté à ce sport populaire traditionnel, avoir vu comment la partie des «batteurs» projette au loin le «frelon» au moyen de la batte flexible et comment la partie des «chasseurs» essaie de le capturer en lançant en l'air les palettes en bois, pour comprendre pleinement le sens de l'affiche: une invitation non verbale à l'achat de billets de la Loterie nationale suisse.

Même pour l'initié, l'affiche révèle souvent foule d'aspects intéressants, parfois essentiels, sur la vie d'un pays, ou laisse entrevoir des détails inattendus, voire inconnus, d'un monde familier que chacun croyait connaître. Cette aptitude à rendre témoignage des conditions de vie et du caractère d'un peuple est typique pour la grande moyenne des affiches conventionnelles, de conception peu différenciée, qui reproduisent leur environnement sans discernement, mais avec d'autant plus de complaisance. Les traces de la vie locale et les caractéristiques de la culture régionale apparaissent toutefois aussi dans les créations peu conven-

superfluous. Thus the same fact — a lack of cinema posters — can theoretically have opposite grounds. The situation becomes more complex when it is taken into account that the custom of publicizing films through newspaper advertising, in a given country, provides a natural explanation for the lack of cinema posters. In Switzerland, there are remarkably few cinema posters apart from those of film clubs and institutes. It would be absurd to draw the conclusion that the Swiss people are not interested in films, or that their attitude is such that all cinema advertising has become superfluous.
The interpretation of specific observations about the poster situation in a given country is therefore not at all easy. Evaluations should produce more valid results, the more the political, economic, social and cultural environment is taken into consideration. Another example illustrates this point:
In recent decades in Switzerland, a remarkably large number of posters have advertised humanitarian works and called for donations. This might lead to the conclusion that, in Switzerland, social security is badly conducted, otherwise there would be no need for such conspicuous appeals for donations to help the elderly, and also the young, the sick and the homeless. In point of fact, social works subsidized by tax-payers are better organized in Switzerland than in most of the surrounding countries. Does Swiss charity therefore reflect the guilty conscience of a prosperous people; or are the donations a form of thanksgiving for the donors' own well-being? One must therefore beware of hasty conclusions from "static" observations.
The posters shown in the present publication are divided into sixteen thematic groups, of which a smaller proportion cover important areas of political life and the majority are concerned with well-known and significant social and humanitarian enterprises. Neither the political nor the social section covers all possible and existing themes, as seen on poster sites. A special characteristic of Swiss posters is that they are very often used for "one-off" political, and especially social, charitable and humanitarian objectives or purposes. Such concentrated efforts have enlivened the poster landscape over the years to a gratifying extent, but it is difficult to include them in collective thematic groups.
The selection is limited to themes which are characteristic of the uses of the poster in Switzerland. Almost all of these themes have had manifold results in poster art and have repeatedly led to convincing, positive solutions by the artists. Where less attention was paid to "social" and "political" themes in posters, it was less possible to regard them as artistically valuable or outstanding. The

Wirksamkeit aus, dass es Tatsachen oder Gefühlsmomente aus einem vergessenen Gestern oder Vorgestern wieder ins Bewusstsein hebt.
Die Einblicke in nationale oder lokale Mentalitäten sind um so genauer und «gehaltvoller», je mehr die Möglichkeit besteht, gleichzeitig eine Anzahl von Plakaten ins Visier zu nehmen, vorzugsweise von gleicher oder ähnlicher Thematik. Hier ein paar – extrem vereinfachte – Beispiele:
Finden sich in der Plakatproduktion eines Landes keine oder fast keine Warenplakate, so liegt die Vermutung nahe, dass das Fehlen einer Wirtschaftswerbung das Fehlen einer freien Marktwirtschaft signalisiert. Wo Planwirtschaften herrschen, bestimmt nicht individueller oder angereizter Bedarf die Produktion und Verteilung der Güter.
Das Fehlen von Filmplakaten könnte signalisieren, dass in einem Land kein nennenswertes Interesse an Filmkunst besteht (ein absurder Schluss); oder dass im Gegenteil der Filmkonsum so verbreitet, so selbstverständlich ist, dass eine Plakatwerbung sich erübrigt. Derselbe Tatbestand – fehlende Filmplakate – kann also, theoretisch, gegensätzliche Voraussetzungen haben. Komplizierter wird die Situation, wenn mitberücksichtigt wird, dass die in einem Land eingebürgerte Werbung für den Film durch Inserate oder wöchentliche Filmprogramme in den Zeitungen das Fehlen des Filmplakats auf natürliche Weise erklärt. In der Schweiz gibt es, abgesehen von Aktivitäten filmkultureller Institutionen, auffallend wenig Filmplakate. Es wäre absurd, daraus ein mangelndes Interesse der Schweizer am Film abzulesen oder auf eine filmkulturelle Situation zu schliessen, die jede Werbung für den Film überflüssig macht.
Die Interpretation bestimmter Beobachtungen über die Plakatsituation in einem Land ist also nicht ganz einfach. Auswertungen dürften um so gültigere Ergebnisse zeitigen, je mehr auch das politische, wirtschaftliche und soziokulturelle Umfeld in die Betrachtungen einbezogen wird. Auch dafür ein Beispiel:
In den letzten Jahrzehnten wirbt in der Schweiz eine auffallend grosse Zahl von Plakaten für humanitäre Hilfswerke bzw. deren Spendenaktionen. Das könnte zum Schluss führen, in diesem Lande sei es schlecht bestellt um die soziale Sicherheit, andernfalls müsste nicht so lauthals für die Alten, aber auch für die Jungen, für Invalide oder Flüchtlinge gesammelt werden. Tatsächlich sind die durch den Steuerzahler alimentierten Sozialwerke in der Schweiz besser ausgerüstet als in den meisten umliegenden Ländern. Spiegelt die Schweizer Spendefreudigkeit also das schlechte Gewissen eines Volkes im Wohlstand, oder ist der gespendete Obolus die Dankesgabe fürs eigene Wohler-

tionnelles réalisées par des affichistes indépendants. En plus de la fonction spécifique qu'elle assume, l'affiche devient alors analyse, interprétation, étude critique du pays d'origine et de l'environnement social. Parfois elle tire son impact précisément des faits concrets ou des éléments affectifs appartenant à un passé plus ou moins lointain, qu'elle fait émerger jusqu'à la conscience de l'homme moderne.
Ces «flashes» sur des mentalités nationales ou locales deviennent plus précis et plus «substantiels» encore s'il existe la possibilité de les comparer à d'autres affiches, donnant de préférence un éclairage différent du même sujet. Voici quelques exemples, fort simplifiés, pour corroborer cette affirmation:
Si, parmi les affiches produites dans un pays donné, pratiquement aucune n'est consacrée à la présentation des produits, cette absence de publicité économique permet de supposer que l'économie de marché est inexistante. Là où règne l'économie dirigée, la production et la distribution des biens ne sont guère déterminées par les besoins individuels ou stimulés du consommateur.
L'absence d'affiches cinématographiques pourrait signaler que, dans un certain pays, il n'existe aucun intérêt notable pour l'art cinématographique (une conclusion absurde!) ou, au contraire, que la consommation généralisée de productions cinématographiques rend toute publicité superflue. Le même état de faits – l'absence d'affiches cinématographiques – peut donc théoriquement s'expliquer en partant d'hypothèses contraires. La situation devient encore plus complexe quand on considère que la publicité cinématographique traditionnelle s'effectue par la publication d'annonces ou de programmes hebdomadaires dans les journaux, ce qui explique l'absence d'affiches cinématographiques de manière naturelle. Abstraction faite de l'activité déployée par certains instituts culturels, il n'existe en Suisse que peu d'affiches cinématographiques. Il serait absurde d'en déduire un manque d'intérêt pour le film suisse ou de prétendre que la situation culturelle sur le plan cinématographique rend superflue toute publicité spécifique pour le film.
L'interprétation des observations faites sur le rôle de l'affiche dans un pays déterminé n'est donc pas toujours facile. L'évaluation des résultats sera d'autant plus valable que l'environnement politique, économique et socio-culturel aura été intégré dans les considérations. Là encore, un exemple:
Depuis plusieurs décennies, un nombre étonnant d'affiches sont consacrées en Suisse à la publicité pour des œuvres humanitaires et leurs collectes de fonds. On pourrait en déduire que la situation

selection here presented should show, once more, that it is possible not only to apply aesthetic standards to posters, but also to ask questions about their content and their place in cultural history. Some fascinating answers are obtained from both lines of questioning.

The primary objective has been to present some central themes drawn from the multitude of social and political questions reflected by impressive posters, and, in addition, to clarify the historical course of their appearance in Swiss poster art. The posters shown can therefore be seen, quite apart from their formal merits or demerits, as eloquent witnesses of changes in Swiss life from about 1920 to the present day: in the nature of society, in the psychology of advertising, in art, and also in politics. Thus they form a panorama of six eventful decades, during which there have been basic changes in many areas, some far from being completed and some only just beginning.

Willy Rotzler

gehen? Vor Kurzschlüssen aus «statistischen Beobachtungen» muss gewarnt werden.

Die in dieser Publikation vorgelegten Plakate sind in sechzehn thematische Gruppen gegliedert; diese sind, zum kleineren Teil, wichtigen Bereichen des politischen Lebens zugehörig und, zum grösseren Teil, bekannten und bedeutenden sozialen und humanitären Unternehmungen. Weder in der Gattung des politischen noch des sozialen Plakats sind alle denkbaren und praktisch vorkommenden, das heisst auf der Plakatwand erscheinenden Themen berücksichtigt. Es gehört zur spezifisch schweizerischen Plakatsituation, dass immer wieder für einmalige politische und vor allem soziale, karitative oder humanitäre Zielsetzungen oder Zwecke das Plakat genutzt wird. Solche punktuellen Stösse mit einmaligen Plakataktionen beleben über die Jahre hin in erfreulichem Mass die Plakatlandschaft, sie lassen sich jedoch nur schwer in zusammenfassende thematische Gruppen einbinden.

Die Auswahl beschränkt sich auf Themenkreise, die für die Nutzung des Plakats in der Schweiz charakteristisch sind. Fast alle diese Themen haben im Plakat einen vielfältigen Niederschlag gefunden und immer wieder zu überzeugenden plakatkünstlerischen Lösungen geführt. Sowenig wie alle der Thematik «sozial» oder «politisch» zuzuordnenden Plakatmotive Berücksichtigung fanden, sowenig konnten alle als gestalterisch wertvoll oder überragend zu bezeichnenden Plakate berücksichtigt werden. Die vorgelegte Auswahl mag, einmal mehr, zeigen, dass an das Plakat sowohl ästhetische Massstäbe angelegt wie inhaltliche oder kulturgeschichtliche Fragen gerichtet werden können. Aus beiden Richtungen liegen faszinierende Antworten vor.

Im Vordergrund stand das Ziel, einige zentrale Themen aus der Vielfalt sozialer und politischer Fragestellungen und Aufgaben im Spiegel einprägsamer Plakate vorzulegen und überdies den zeitlichen Ablauf ihres Auftretens im schweizerischen Plakatschaffen deutlich zu machen. Die gezeigten Beispiele werden dadurch, ungeachtet ihrer formalen Qualitäten oder Mängel, zu beredten Zeugen gesellschaftlicher, werbepsychologischer, künstlerischer, aber auch politischer Wandlungen in der Schweiz von etwa 1920 bis heute. Ein Panorama also von sechs ereignisreichen Jahrzehnten, in denen sich grundlegende, vielfach noch längst nicht abgeschlossene, ja zum Teil erst angebahnte Veränderungen ergeben haben.

Willy Rotzler

sociale dans ce pays est fort précaire, sans quoi il ne serait pas nécessaire de propager à grands cris ces campagnes de collecte pour les vieux, les jeunes, les invalides ou les réfugiés. En réalité, les œuvres sociales suisses, alimentées par les contribuables de ce pays, sont mieux dotées que les institutions sociales de la plupart des pays avoisinants. La générosité suisse reflète-t-elle donc la mauvaise conscience d'un peuple qui vit dans la prospérité, ou bien l'obole versée est-elle un geste de gratitude pour le bien-être général?

Les affiches montrées dans la présente publication sont subdivisées en seize groupes thématiques, relevant des secteurs importants de la vie politique et, en majeure partie, d'entreprises sociales et humanitaires importantes et bien connues. Ni dans la catégorie de l'affiche politique, ni dans celle de l'affiche sociale, il n'a été possible de présenter tous les thèmes concevables ou effectivement réalisés et placardés sur les murs d'affichage.

Le choix se limite aux catégories de thèmes caractéristiques pour le mode d'utilisation de l'affiche en Suisse. Presque tous ces sujets ont trouvé de multiples formes d'expression et ont abouti à des solutions graphiques convaincantes par l'affiche. Tout comme il a été indispensable d'effectuer un choix parmi tous les motifs à caractère «politique» ou «social», de même il s'est avéré impossible de tenir compte de toutes les affiches de haute qualité formelle et de conception graphique exceptionnelle. Le choix présenté révèle, une fois de plus, que l'affiche peut répondre tant à des critères esthétiques qu'à des questions de fond relatives aux problèmes de la vie moderne ou de l'histoire culturelle. Des réponses fascinantes ont été données dans les deux sens.

L'objectif primaire consistait à présenter quelques thèmes centraux, tirés de la vaste multiplicité des tâches et questions sociales et politiques – sous forme de «flash direct» sur des affiches suggestives – et d'illustrer l'ordre chronologique de leur apparition sur la scène de l'affiche en Suisse. Indépendamment de leurs qualités ou défauts formels, les exemples cités deviennent le témoignage éloquent des changements intervenus en Suisse de 1920 environ jusqu'à nos jours, à la fois sur le plan social, psychologique et publicitaire, artistique, et aussi politique. Ils forment autant de facettes d'un vaste panorama qui s'ouvre sur six décennnies lourdes en événements variés et marquées de mutations profondes, dont certaines sont encore en cours et d'autres viennent tout juste d'être amorcées.

Willy Rotzler

Insurance for the Elderly and Bereaved

One of the worthiest concerns of the democratic state is to take over social obligations towards the weak and the handicapped. A rule which should be respected in such welfare activities is that the beneficiaries should not be given the impression that they are recipients of charity. The aid must be given on the basis of a legal right to which anyone may become entitled.
Ardent campaigning on behalf of the various claims of welfare for the old, the bereaved and the sick began in Switzerland in the 1920s. The plebiscites of 1925, 1931 and 1947 were major events in these democratic controversies. Although the social services concerned have long since been put into effect, and however much we may now take them for granted, it should not be forgotten that, at the time of the plebiscites, the matters under discussion were not only of a practical nature but were also profoundly connected with a question of national politics: Should the democratic state turn itself into a Welfare State?
Each of the plebiscites also brought forth an artistic contest on the poster hoardings. The social theme made the work of interest to artists who took an interest in the human condition. Thus it happened that, as well as the poster designer Carl Scherer, whose political engagement was especially marked, others such as the painter and muralist Alfred Heinrich Pellegrini of Basle and the landscape artist Hans Beat Wieland stated their views through powerful posters. In the decisive plebiscite of 1947, which brought about the definitive change, Hans Erni found the best artistic solution with the bold form and style of his poster showing old and young people. His contribution was all the more effective in that he intensified the theme with both male and female subjects. Swiss voters accepted the important social welfare proposals by an overwhelming majority on 6th June, 1947.

Kampf um Alters- und Hinterbliebenen-Versicherung

Es gehört zu den vornehmsten Anliegen des demokratischen Staates, soziale Verpflichtungen gegenüber den Schwachen und Benachteiligten zu übernehmen. Dabei gilt als Spielregel dieser sozialen Hilfsmassnahmen, bei den Nutzniessern nicht den Eindruck aufkommen zu lassen, sie seien Almosen-Empfänger. Die Hilfe muss auf der Basis eines Rechtsanspruches erfolgen, den grundsätzlich jeder geltend machen könnte.
In der Schweiz ist seit den zwanziger Jahren um die verschiedenen Postulate von Alters-, Hinterbliebenen- und Invalidenversicherung leidenschaftlich gerungen worden. Grosse Volksabstimmungen von 1925, 1931 und 1947 sind Hauptstationen dieser demokratischen Auseinandersetzungen. So selbstverständlich uns heute die inzwischen längst realisierten Sozialwerke vorkommen, darf man doch nicht vergessen, dass in der Zeit, in der sie diskutiert wurden, es nicht nur um Sachgeschäfte, sondern auch um staatspolitische Grundsatzfragen ging: Soll der demokratische Staat zu einem eigentlichen Wohlfahrtsstaat werden, hiess eine dieser Fragen.
Jede der genannten Abstimmungen löste auch einen künstlerischen Wettstreit auf den Plakatwänden aus. Die soziale Thematik machte die Aufgabe vor allem für solche Künstler interessant, deren Aufmerksamkeit der «Condition humaine» galt. So fällt auf, dass neben dem politisch besonders engagierten Plakatgraphiker Carl Scherer der Basler Maler und Wandbildmaler Alfred Heinrich Pellegrini oder der Landschafter Hans Beat Wieland sich mit packenden Plakaten zum Wort meldeten. In der entscheidenden AHV-Abstimmung von 1947, die den endgültigen Durchbruch brachte, hat Hans Erni mit seiner formal und stilistisch kühnen Kombination des alten und des jungen Menschen die plakatkünstlerisch gültigste Lösung gefunden. Ernis Beitrag war um so wirkungsvoller, als er das Thema mit einer weiblichen Variante intensivierte. Mit einem überwältigenden Mehr haben die Schweizer Stimmbürger am 6. Juni 1947 das grosse Sozialwerk angenommen.

Lutte pour l'assurance-vieillesse et survivants

L'intervention en faveur des faibles et défavorisés constitue l'une des obligations les plus nobles de l'Etat démocratique. Il est toutefois essentiel que ces mesures d'aide sociale ne suscitent pas chez le bénéficiaire l'impression de recevoir des aumônes. L'assistance doit intervenir sur la base d'un droit légitime que chacun peut en principe faire valoir.
En Suisse, les postulats relatifs à l'assurance-vieillesse, survivants et invalidité ont été l'objet de discussions passionnées depuis les années vingt. Les votations populaires des années 1925, 1931 et 1947 constituent des jalons importants dans cette confrontation démocratique de grande envergure. Aujourd'hui, ces œuvres sociales sont pour nous une réalité évidente; mais n'oublions pas qu'à l'époque où elles étaient en discussion, ces questions spécifiques impliquaient des options beaucoup plus fondamentales encore, telles que: notre Etat démocratique doit-il devenir un Etat-providence?
Chacune des votations mentionnées déclencha sur les murs d'affichage une véritable compétition artistique. Les thèmes sociaux stimulaient spécialement les artistes intéressés par la «condition humaine». Aussi ne faut-il guère s'étonner de voir les affiches politiques très engagées du graphiste Carl Scherrer à côté des œuvres percutantes du Bâlois Alfred Heinrich Pellegrini, peintre et réalisateur de fresques, ou du paysagiste Hans Beat Wieland. Pour la votation AVS en 1947, qui a marqué la percée décisive, Hans Erni a trouvé une solution des plus éloquentes: il présente l'homme âgé associé à l'homme encore jeune en une combinaison stylistique et formelle audacieuse. La contribution d'Erni est d'autant plus suggestive qu'il a encore intensifié le thème par une variante féminine. Avec une majorité écrasante de voix, les citoyens suisses ont accepté, le 6 juin 1947, la grande œuvre sociale de l'AVS.

1.

Karl Bickel
1925, 90,5 × 128 cm
Lithography / Lithographie / Lithographie

2.

Alfred Heinrich Pellegrini
1925, 90,5 × 128 cm
Lithography / Lithographie / Lithographie

3.

Carl Scherer
1925, 90,5 × 128 cm
Lithography / Lithographie / Lithographie

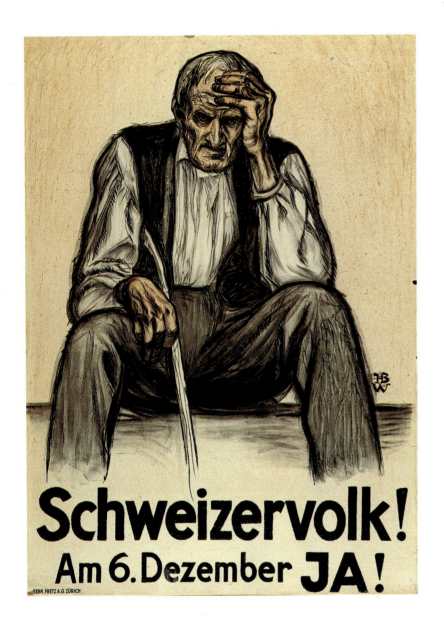

4.

Hans Beat Wieland
1931, 90,5 × 128 cm
Lithography / Lithographie / Lithographie

5.

Carl Scherer
1931, 90,5 × 128 cm
Lithography / Lithographie / Lithographie

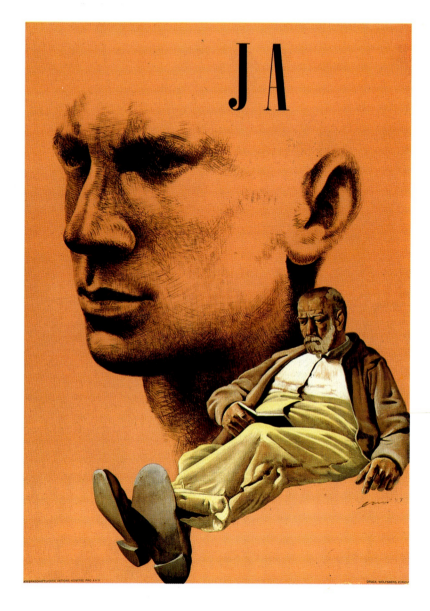

6.

Carl Scherer
1931, 90,5 × 128 cm
Draft of poster / Plakatentwurf / Projet d'affiche

7.

Hans Erni
1947, 90,5 × 128 cm
Lithography / Lithographie / Lithographie

8.

Anonym
1947, 90,5 × 128 cm
Lithography / Lithographie / Lithographie

9.

Anonym
1947, 90,5 × 128 cm
Lithography / Lithographie / Lithographie

10.

Walter Monticelli
1947, 90,5 × 128 cm
Lithography / Lithographie / Lithographie

11.

Anonym
1972, 90,5 × 128 cm
Silkprint / Siebdruck / Sérigraphie

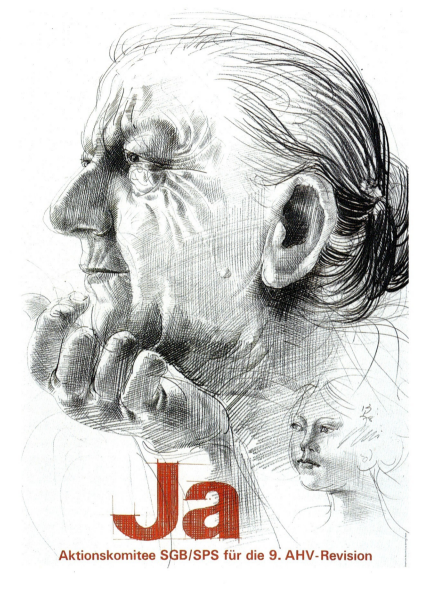

12.

R. Staub
1972, 90,5 × 128 cm
Offset printing / Offsetdruck / Impression offset

13.

Hans Erni
1978, 90,5 × 128 cm
Offset printing / Offsetdruck / Impression offset

Women's Suffrage

Foreigners have long seen it as something of a joke, that women were not included in the alleged general equality of the oldest democracy. They forget, however, that in most countries the introduction of women's voting rights was imposed by government authority, not by a decision of the people. In Switzerland, no political subject of national importance has ever been disputed so long, so ardently or with such frequent use of unpleasant tactics, as that of women's suffrage.

The poster, as a mirror of these controversies, is extremely revealing, not least in a psychological sense. The first plebiscite on women's suffrage was held in 1920 — certainly under the pressure of world war and major social upheavals, including the general strike of 1918. While the opponents warned of "bluestocking rule", the supporters of women's voting rights convincingly gave prominence to human solidarity, to woman's maternal role and also to professional activities such as nursing, though mostly through illustration rather than through poster art in the true sense.

A further plebiscite in 1946 gave rise to interesting poster themes and forms, such as Donald Brun's "comforter". It is noteworthy that in some cases the same theme, the child, was used in advertising both for and against women's suffrage; and also that many feeble and pointless posters, often anonymous, came to be displayed. Once again, Hans Erni made an artistically valid contribution with a poster showing specific qualities of draughtsmanship and illustration.

The plebiscites then become more frequent. In 1959, Celestino Piatti supplies the strongest approach with a powerful poster on the theme of chivalry. The subsequent plebiscites of 1966 and 1971 (the latter finally introducing women's suffrage at the federal level) are remarkably unproductive from the point of view of poster art. Inspiring themes and visualizations are lacking. The Yes-vote of the majority was certainly not stimulated by any outstanding posters.

Das Frauenstimmrecht

Ausländer haben sich lange über die Schweiz lustig gemacht, weil in der ältesten Demokratie die angebliche Gleichheit aller die Frauen nicht mit einschloss. Dabei ist allerdings vergessen worden, dass in den meisten Ländern die Einführung des Frauenstimmrechts nicht durch Volksentscheide zustande kam, sondern von den Regierungen dekretiert worden ist. In der Schweiz hat es kein politisches Thema von landesweiter Bedeutung gegeben, um das so lange, so leidenschaftlich und oft auch mit so unschönen Mitteln gekämpft worden ist wie ums Frauenstimmrecht.

Das Plakat als Spiegel dieser Auseinandersetzung ist, wohl auch psychologisch, recht aufschlussreich: 1920 fand – gewiss unter dem Eindruck des Weltkriegs und der grossen sozialen Erschütterungen (Generalstreik von 1918) – die erste Abstimmung übers Frauenstimmrecht statt. Während die Gegner auch in der Plakatwerbung vor dem Blaustrumpf warnten, stellten die Befürworter überzeugend die Solidarität der Geschlechter, die Frau in ihrer Rolle als Mutter, die Frau aber auch in den helfenden Berufen, etwa als Krankenschwester, in den Vordergrund. Dies alles meist eher mit illustrativen als mit plakativen Mitteln.

Eine weitere Abstimmung von 1946 zeitigte motivisch und formal interessante Plakatlösungen, etwa den Schnuller von Donald Brun. Bemerkenswert ist, wie teils mit dem gleichen Motiv, dem Kind, für und gegen das Frauenstimmrecht geworben wurde. Ebenso bemerkenswert ist, wie viele belanglose, oft auch anonyme Plakate ohne Durchschlagskraft zum Aushang kamen. Einmal mehr lieferte Hans Erni mit einer – spezifisch zeichnerisch-illustrativen– Lösung einen künstlerisch gültigen Beitrag.

Dann häufen sich die Abstimmungen. 1959 bringt Celestino Piatti in einem kraftvollen Plakat mit dem Motiv der Ritterlichkeit die stärkste Lösung. Die folgenden Abstimmungen von 1966 und 1971 (die letzte bringt auf Bundesebene endlich das Frauenstimmrecht) sind plakatkünstlerisch merkwürdig wenig ergiebig. Zündende Motive und Visualisierungen blieben aus. Das Ja der Mehrheit ist sicher nicht durch glanzvolle Plakate stimuliert worden.

Le droit de vote des femmes

Les étrangers ont longtemps cité la Suisse d'un petit air narquois en rappelant que dans cette plus ancienne démocratie, la prétendue égalité pour tous n'incluait pas les femmes. Or, dans la plupart des pays, le droit de vote des femmes n'a pas été introduit en vertu de décisions populaires, mais par décret gouvernemental. En Suisse, aucun autre sujet politique de portée nationale n'a fait l'objet d'autant de débats passionnés et d'expédients aussi contestables que le suffrage féminin.

L'affiche, en tant que reflet de ces confrontations, révèle le climat général et les arguments psychologiques du débat. Encore sous l'influence de la Première Guerre mondiale et des grands bouleversements sociaux (grève générale de 1918), la première votation populaire sur le droit de vote des femmes eut lieu en 1920. Tandis que les adversaires mettaient en garde contre les bas-bleus, les promoteurs évoquaient avec force la solidarité entre les sexes, la femme dans son rôle de mère, et aussi sa fonction dans les professions soignantes. Cette propagande se servait de préférence de moyens illustratifs plutôt que de ceux de l'affiche. La votation de 1946 suscita des affiches intéressantes, tant du point de vue des motifs que de la conception formelle, comme par exemple la sucette de Donald Brun. Relevons que parfois le même motif – l'enfant – était utilisé pour militer tantôt pour et tantôt contre le suffrage féminin. Il est intéressant aussi de constater combien d'œuvres sans intérêt, parfois même anonymes et dénuées de tout effet suggestif, ont été placardées sur les murs. Une fois de plus, Hans Erni a fourni une contribution artistique déterminante avec sa création picturale spécifiquement illustrative.

Puis les votations se succédèrent. En 1959, Celestino Piatti a créé la solution la plus convaincante avec une affiche suggestive, empreinte du motif de la courtoisie. Les votations ultérieures (en 1966 et en 1971, la dernière aboutissant à l'institution du suffrage féminin sur le plan fédéral) n'ont produit que peu d'affiches marquantes. Les motifs et visualisations percutantes font défaut. Le oui de la majorité n'a guère été obtenu grâce à des affiches particulièrement brillantes.

14.

Otto Baumberger
1920, 90,5 × 128 cm
Lithography / Lithographie / Lithographie

15.

Margrit Gams
1920, 90,5 × 128 cm
Lithography / Lithographie / Lithographie

16.

Dora Hauth
1920, 90,5 × 128 cm
Lithography / Lithographie / Lithographie

17.

Dora Hauth
1920, 80×116 cm
Lithography / Lithographie / Lithographie

18.

Alfred Heinrich Pellegrini
1920, 90,5×128 cm
Lithography / Lithographie / Lithographie

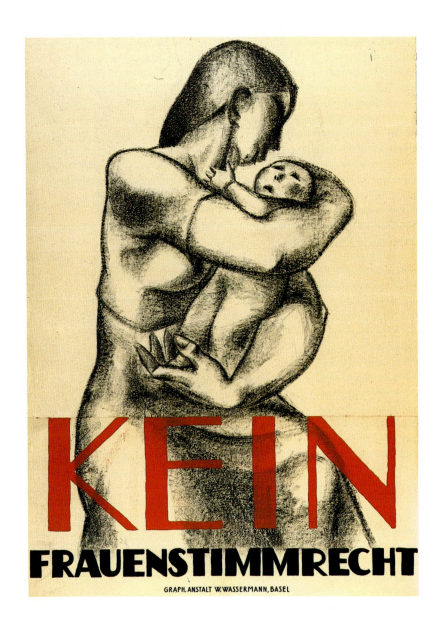

19.

Niklaus Stoecklin
1920, 90,5 × 128 cm
Lithography / Lithographie / Lithographie

20.

Ernst Keiser
1920, 90,5 × 128 cm
Lithography / Lithographie / Lithographie

21.

Anonym
1946, 90,5 × 128 cm
Lithography / Lithographie / Lithographie

22.

Donald Brun
1946, 90,5 × 128 cm
Lithography / Lithographie / Lithographie

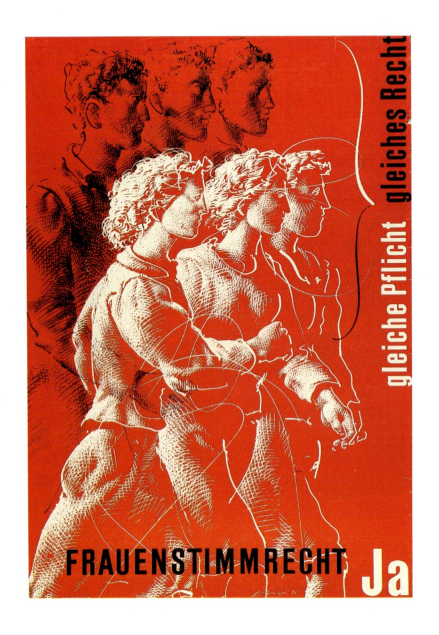

23.

Hans Erni
1946, 90,5 × 128 cm
Lithography / Lithographie / Lithographie

24.

Hugo Laubi
1946, 90,5 × 128 cm
Lithography / Lithographie / Lithographie

25.

Beatrice Afflerbach
1946, 90,5 × 128 cm
Silkprint / Siebdruck / Sérigraphie

26.

Beatrice Afflerbach
1946, 90,5 × 128 cm
Lithography / Lithographie / Lithographie

27.

Anonym
1946, 90,5 × 128 cm
Lithography / Lithographie / Lithographie

28.

Noël Fontanet
1946, 90,5 × 128 cm
Lithography / Lithographie / Lithographie

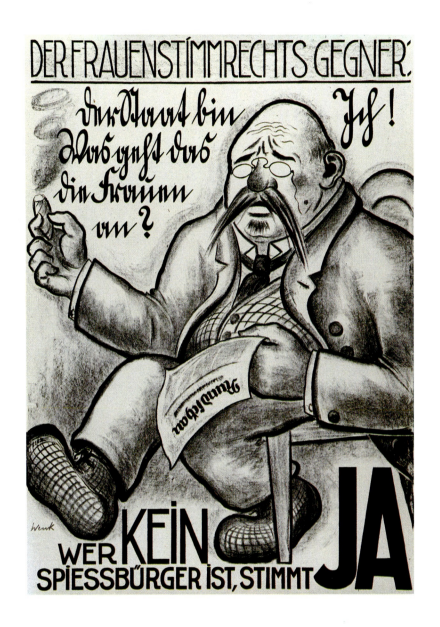

29.

Wilhelm Wenk
1946, 90,5 × 128 cm
Lithography / Lithographie / Lithographie

30.

Anonym
1954, 90,5 × 128 cm
Lithography / Lithographie / Lithographie

31.

Jürg Spahr
1959, 90,5 × 128 cm
Lithography / Lithographie / Lithographie

32.

Celestino Piatti
1959, 90,5 × 128 cm
Offset printing / Offsetdruck / Impression offset

33.

Donald Brun
1959, 90,5×128 cm
Lithography / Lithographie / Lithographie

34.

Anonym
1959, 90,5×128 cm
Offset printing / Offsetdruck / Impression offset

35.

René Gilsi
1959, 90,5 × 128 cm
Lithography / Lithographie / Lithographie

36.

Anonym
1959, 90,5 × 128 cm
Lithography / Lithographie / Lithographie

37.
Anonym
1966, 90,5×128 cm
Offset printing / Offsetdruck / Impression offset

38.
Anonym
1966, 90,5×128 cm
Silkprint / Siebdruck / Sérigraphie

39.
Buchdruckerei Baden AG
1971, 90,5 × 128 cm
Offset printing / Offsetdruck / Impression offset

40.
Anonym
1971, 90,5 × 128 cm
Offset printing / Offsetdruck / Impression offset

41.

Peter Freis
1971, 90,5 × 128 cm
Offset printing / Offsetdruck / Impression offset

42.

Anonym
1971, 90,5 × 128 cm
Offset printing / Offsetdruck / Impression offset

Shorter Working Hours

The limitation of working hours to a physically and mentally bearable maximum is one of the oldest demands of the working class. Whereas the 60-hour week was once a worthy objective, our times (from 1920 onwards) have seen struggles for the 44-hour week. And yet the opposite has also been known: as is proved by posters, a revision of the Factory Law in Switzerland came up for discussion in 1924, with the objective of lengthening working hours!
A renewed battle for the general application of the 44-hour week in 1958 yielded two effective Yes posters: one by Heiner Bauer with spirited qualities of painting and an industrial landscape by the political artist René Gilsi, advocating the liberation of workers for two days of freedom after "five days for daily bread" in the world of cranes, factory chimneys and high-voltage cables.
In 1976, in a climate of economic recession and unemployment, an initiative was launched by the New Left for a 40-hour week, regarded as a means of creating new jobs. From a series of anonymous and rather formless posters, there emerges a well designed and also typographically convincing photo-poster by Bernard Schlup, which provides by far the best answer and, in its selection and use of pictorial material, is in the true sense a "political poster".
Swiss demands have not subsequently gone below the 40-hour level, but it is to be expected that, depending on the economic situation, efforts to reduce working hours still further will continue, as in other countries.

Kürzere Arbeitszeit

Die Begrenzung der Arbeitszeit auf ein physisch und psychisch erträgliches Mass gehört zu den ältesten Forderungen der Arbeiterklasse. War einst die 60-Stunden-Woche ein erstrebenswertes Ziel, so ist in unserem Zeitalter, von 1920 an, um die 44-Stunden-Woche gekämpft worden. Doch auch das Umgekehrte gab es: Wie Plakate belegen, stand im Jahre 1924 in der Schweiz eine Revision des Fabrikgesetzes zur Diskussion, die eine Verlängerung der Arbeitszeit gebracht hätte!
Ein erneuter Kampf um die allgemeinverbindliche Festlegung der 44-Stunden-Woche im Jahre 1958 brachte als graphischen Ertrag zwei wirkungsvolle Ja-Plakate: ein malerisch-stimmungshaftes von Heiner Bauer und eine Industrie-Landschaft des politischen Zeichners René Gilsi, der den Arbeitenden aus der Welt der Krane, Fabrikschlote und Hochspannungsleitungen nach «fünf Tagen fürs Brot» für zwei Tage in die Freiheit entfliegen lässt.
1976 taucht im Klima der wirtschaftlichen Rezession und der Arbeitslosigkeit eine von der jungen Linken lancierte Initiative für eine 40-Stunden-Woche auf, gedacht als Mittel zur Beschaffung zusätzlicher Arbeitsplätze. Aus einer Reihe von anonymen, formal wenig artikulierten Ja-Plakaten ragt das gut gebaute, auch in der Schrift überzeugende Photoplakat von Bernard Schlup als wohl stärkste Lösung heraus – in der Wahl und Anwendung der bildnerischen Mittel im wahren Sinn ein «politisches Plakat».
Unter die 40-Stunden-Marke sind seitdem in der Schweiz die Forderungen nicht gegangen. Doch ist zu erwarten, dass, je nach der wirtschaftlichen Lage, wie in andern Ländern die Bemühungen um weitere Arbeitszeit-Verkürzungen weitergehen werden.

La réduction de la durée du travail

La réduction de l'horaire de travail constitue une des plus anciennes revendications de la classe ouvrière pour aboutir à une durée physiquement et psychiquement supportable. Alors que jadis la semaine de 60 heures paraissait un objectif souhaitable, l'horaire hebdomadaire de 44 heures a été revendiqué dès 1920. Mais l'inverse s'est également produit: en 1924, une révision de la loi sur les fabriques prévoyait un prolongement de la durée du travail. Les affiches témoignent de cette discussion.
Une nouvelle lutte s'engagea en 1958 autour du caractère dit «de force obligatoire» de la semaine de 44 heures. Elle aboutit à deux créations graphiques percutantes, militant pour un oui: une affiche picturale très évocatrice de Heiner Bauer et une œuvre politique du dessinateur René Gilsi, montrant un paysage industriel de grues, cheminées d'usine et mâts de haute tension d'où le travailleur, après «cinq jours pour gagner son pain», s'échappe pour jouir de deux jours de liberté.
En 1976, dans un climat de récession économique et de chômage, la jeune gauche lança l'initiative pour la semaine de 40 heures, conçue comme moyen de créer des emplois supplémentaires. Parmi toute une série d'affiches anonymes, plaidant pour un oui au moyen d'une conception formelle souvent peu convaincante, l'affiche photographique de Bernard Schlup se détache avec force par sa conception formelle et typographique bien articulée, ainsi que par le choix et l'application des moyens créatifs. C'est une «affiche politique» au vrai sens du mot.
Depuis lors, les revendications en Suisse n'ont guère encore été en-dessous de la semaine de 40 heures. En considération de la situation économique générale et des revendications dans les autres pays, il faut toutefois s'attendre à de nouvelles demandes de réduction de la durée du travail.

43.

Paul Kammüller
1920, 90,5 × 128 cm
Lithography / Lithographie / Lithographie

44.

Carl Scherer
1920, 90,5 × 128 cm
Lithography / Lithographie / Lithographie

45.

Carl Scherer
1920, 90,5 × 128 cm
Lithography / Lithographie / Lithographie

46.

Paul Kammüller
1920, 90,5 × 128 cm
Lithography / Lithographie / Lithographie

47.

Paul Wyss
1920, 90,5 × 128 cm
Lithography / Lithographie / Lithographie

48.

Jules Courvoisier
1924, 90,5 × 128 cm
Lithography / Lithographie / Lithographie

49.

Anonym
1924, 24,5 × 32,5 cm
Lithography / Lithographie / Lithographie

50.

Alfred Heinrich Pellegrini
1924, 90,5 × 128 cm
Lithography / Lithographie / Lithographie

51.

Florentin Moll
1924, 90,5 × 128 cm
Lithography / Lithographie / Lithographie

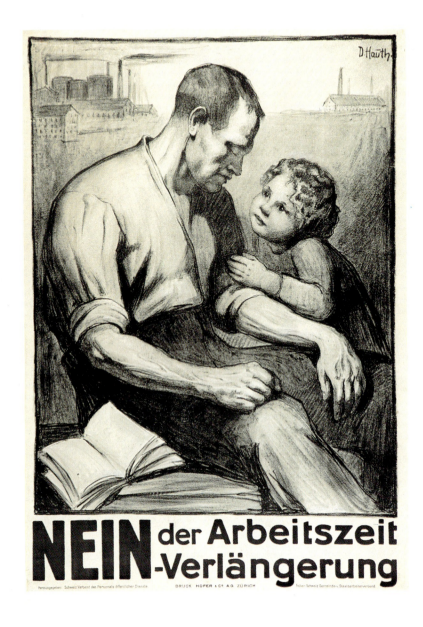

52.

Dora Hauth
1924, 81×115 cm
Lithography / Lithographie / Lithographie

53.

Anonym
1924, 90,5×128 cm
Lino cut / Linolschnitt / Gravure sur linoléum

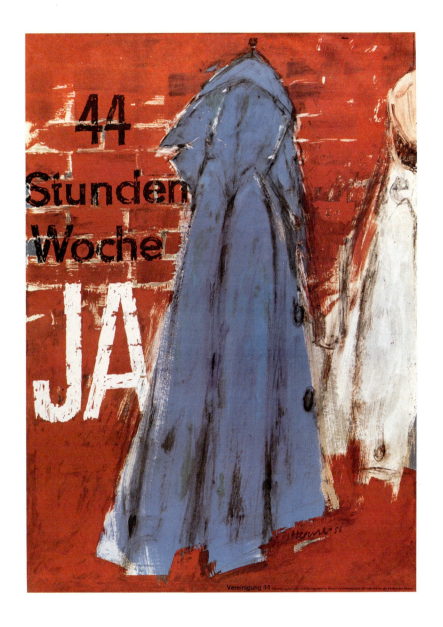

54.

Heiner Bauer
1958, 90,5 × 128 cm
Offset printing / Offsetdruck / Impression offset

55.

René Gilsi
1958, 90,5 × 128 cm
Lithography / Lithographie / Lithographie

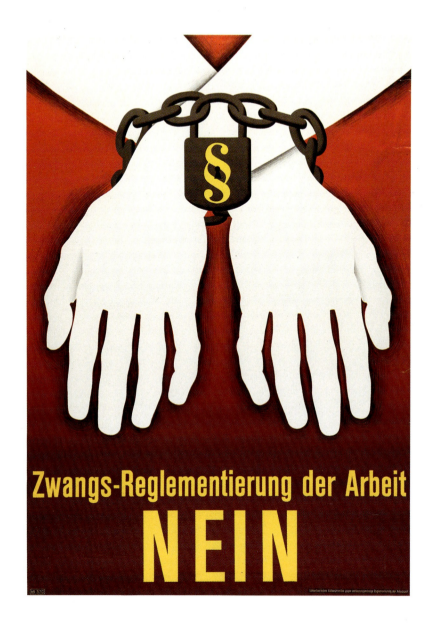
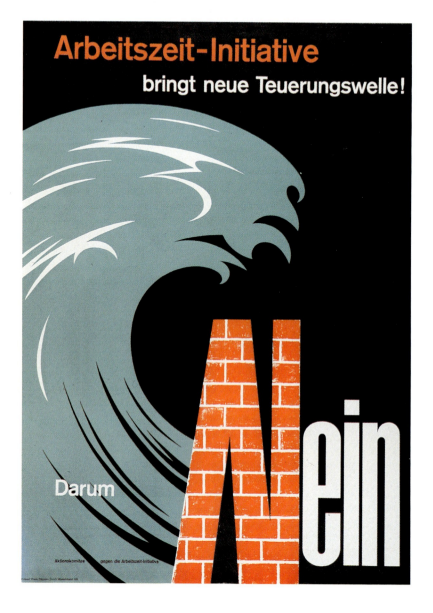

56.

Anonym
1958, 90,5 × 128 cm
Offset printing / Offsetdruck / Impression offset

57.

Erwin Däppen
1958, 90,5 × 128 cm
Offset printing / Offsetdruck / Impression offset

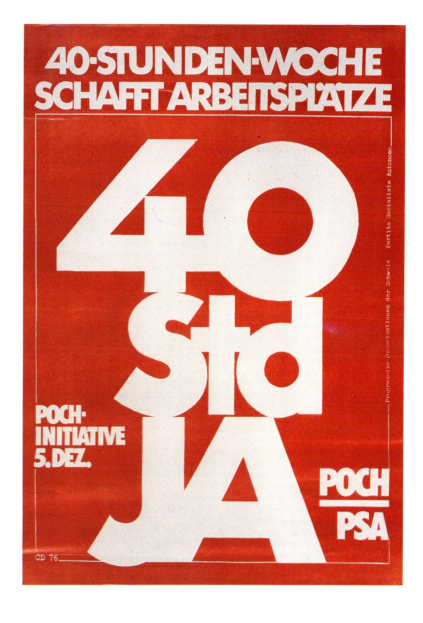

58.

Bernard Schlup
1976, 90,5 × 128 cm
Silkprint / Siebdruck / Sérigraphie

59.

Signatur CD
1976, 41 × 58 cm
Offset printing / Offsetdruck / Impression offset

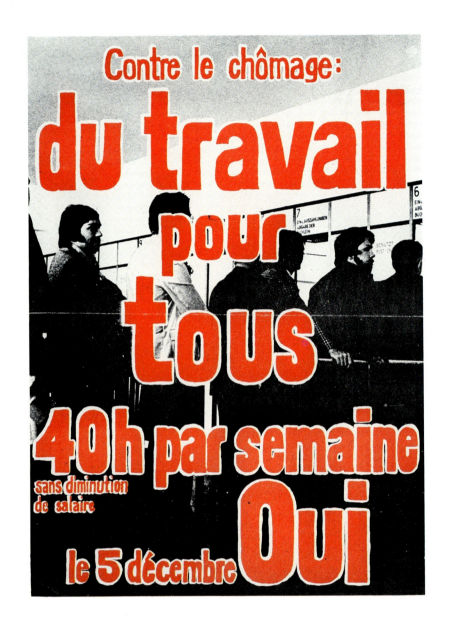

60.

Anonym
1976, 30×42 cm
Offset printing / Offsetdruck / Impression offset

61.

Anonym
1976, 43,5×61 cm
Offset printing / Offsetdruck / Impression offset

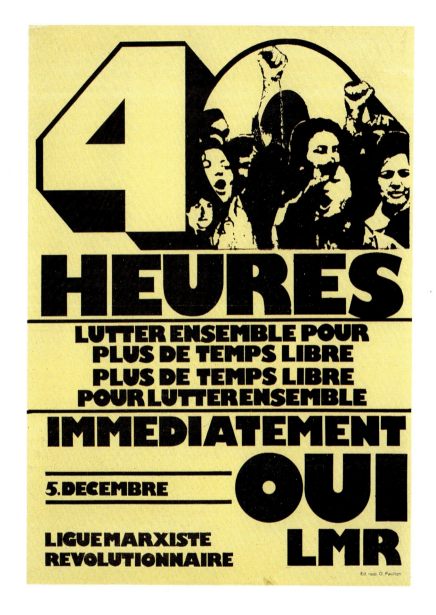

62.

Anonym
1976, 30,5 × 43 cm
Offset printing / Offsetdruck / Impression offset

63.

Anonym
1976, 32 × 46 cm
Offset printing / Offsetdruck / Impression offset

Atomic Defence

The very different questions of atomic war and the peaceful use of atomic power have never been clearly separable, ever since they became matters of general discussion. There is no longer any question of equipping the Swiss Army with atomic weapons, particularly not since the movement against atomic armament became active around 1960. The effectiveness of that campaign against atomic weapons, which distinguished the peace movement in the late 1950s, was partly due to a poster by Hans Erni with an unequalled warning effect. From a death's head, dramatically placed on a black background, a mushroom cloud rises, producing a surrealistic interpenetration of cause and effect.

From the start, an object of genuine democratic controversy was the question whether we should increasingly produce our energy through atomic power stations, or whether we should do so at all. This controversy, in which we take sides today with regard to the living conditions of future generations, is still in full swing. If, at first, a population with a belief in industrial growth and an orientation towards that growth regarded the protestors against atomic power stations as "crackpots", today a reduced number of supporters faces an increasing number of respected opponents.

The outcome of this discussion is uncertain. The question is not so much about matters of fact as, in the long run, about the principle of how far the people have a right to participate in decisions on questions with such grave consequences. This became apparent in 1979 with the controversy about the atomic defence initiative, reflected in a large number of powerful posters, mostly using the characteristic form of the cooling tower as the main theme, combined in many cases with the William Tell motif. This poster campaign involved fewer graphic designers than politically engaged painters (Ruedi Baumann, Pierre Brauchli, Hans Gantert, Hugo Schuhmacher and others). Their aim was to illustrate the possible dangers of even the peaceful use of atomic power. Seldom has the subject of a plebiscite given rise to so many strongly effective, and at the same time artistically remarkable posters. This important question called for serious and intensive interpretations.

Atomschutz

Die als Themen sehr verschiedenartigen Fragen Atomkrieg und friedliche Nutzung der Atomkraft haben sich, seit sie zum Gesprächsstoff wurden, nie klar trennen lassen. Eine Ausrüstung der Schweizer Armee mit Atomwaffen steht heute nicht zur Diskussion, vor allem nicht seit der um 1960 aktiven Bewegung gegen die atomare Aufrüstung. Jenem Kampf gegen die Atomwaffen, wie er in den späten fünfziger Jahren die Friedensbewegungen ausgezeichnet hat, verdanken wir ein in seiner Wirkung unerreichtes warnendes Plakat von Hans Erni: Aus einem dramatisch vor schwarzem Grund aufscheinenden Totenkopf wächst ein Atompilz empor; eine surreale Durchdringung von Ursache und Wirkung.

Ein Gegenstand echter demokratischer Auseinandersetzung war von Anfang an die Frage, ob wir unseren Energiebedarf überhaupt oder vermehrt mit Atomkraftwerken decken wollen. Diese Auseinandersetzung, in der wir heute zu den Lebensbedingungen künftiger Generationen Stellung beziehen, ist noch voll im Gang. Waren es anfänglich in den Augen einer wachstumsorientierten, wachstumsgläubigen Bevölkerung sektiererische «Spinner», die gegen die Atomkraftwerke protestierten, so stehen sich heute eine wachsende Zahl von – respektierten – Gegnern und eine dank zunehmender Skepsis verminderte Zahl von Befürwortern gegenüber.

Der Ausgang dieser Diskussion ist ungewiss. Es geht bei ihr ja nicht nur um Sachfragen, sondern letztlich darum, wieweit in so folgenschweren Fragen das Volk ein Mitspracherecht hat. Das wurde 1979 mit der Auseinandersetzung um die Atomschutz-Initiative deutlich. Sie spiegelte sich in einer grossen Zahl starker Plakate, die zumeist die charakteristische Form des Kühlturms zum Hauptmotiv machten, vielfach anzüglich kombiniert mit dem Tellenmotiv. An dieser Plakataktion waren weniger graphische Gestalter beteiligt als engagierte Maler (Ruedi Baumann, Pierre Brauchli, Hans Gantert, Hugo Schuhmacher u.a.). Ihr Ziel war es, die möglichen Gefährdungen auch durch friedliche Nutzung der Atomkraft zu veranschaulichen. Selten hat ein Abstimmungsthema so viele wirkungsstarke und zugleich künstlerisch bemerkenswerte Plakate gezeigt. Die bedeutungsvolle Frage rief nach ernsthafter und intensiver Interpretation.

La protection contre les dangers atomiques

Les questions relatives à la guerre atomique et à l'utilisation pacifique de l'énergie nucléaire n'ont jamais été clairement distinguées dans les débats publics, bien que les deux thèmes soient de nature foncièrement différente. L'équipement de l'armée suisse en armes atomiques n'est plus guère matière à discussion de nos jours, en particulier depuis l'intervention active du Mouvement contre le réarmement atomique en 1960. C'est à cette lutte contre les armes nucléaires, menée par les mouvements pacifistes vers la fin des années cinquante, que nous devons une affiche incisive de Hans Erni. Le message d'avertissement produit un effet dramatique inégalé. A partir d'une tête de mort dressée sur fond noir se développe une gigantesque nuée radioactive en forme de champignon: une interprétation surréaliste de la relation de cause à effet.

Dès le départ, le véritable débat démocratique visait à établir si, en principe, nous voulons couvrir toujours davantage nos besoins énergétiques par des centrales nucléaires. Cette confrontation, si déterminante pour les conditions de vie des générations futures, est encore en cours. Alors qu'à l'origine, l'orientation vers la croissance comme «source de tout progrès» incitait la population à qualifier les opposants aux centrales nucléaires de «fous» sectaires, il existe aujourd'hui un scepticisme généralisé face au «nucléaire», avec un nombre toujours croissant d'adversaires «respectables» et des protagonistes toujours moins nombreux.

L'issue de cette discussion est incertaine. En fait, il ne s'agit pas primairement de trancher ces questions spécifiques, mais de savoir dans quelle mesure le peuple suisse est admis à participer à des décisions sur des problèmes ayant des répercussions aussi graves. L'initiative antiatomique de 1979 révèle très clairement cet aspect. Toute une série d'affiches percutantes, ayant pour principal motif la forme caractéristique des tours de refroidissement, souvent dans une combinaison ambivalente avec le motif de Tell, témoignent de cette confrontation. Leurs auteurs étaient d'ailleurs beaucoup moins des créateurs graphiques que des peintres engagés (Ruedi Baumann, Pierre Brauchli, Hans Gantert, Hugo Schuhmacher et d'autres). Rarement une question soumise au vote avait-elle produit autant d'affiches aussi éloquentes et de si haute qualité artistique. L'importance du sujet exigeait une interprétation sérieuse et intense.

64.

Hans Erni
1954, 90,5 × 128 cm
Offset printing / Offsetdruck / Impression offset

65.

Cioma Schönhaus
1963, 90,5 × 128 cm
Offset printing / Offsetdruck / Impression offset

66.

Peter Hürzeler
1979, 42 × 60 cm
Silkprint / Siebdruck / Sérigraphie

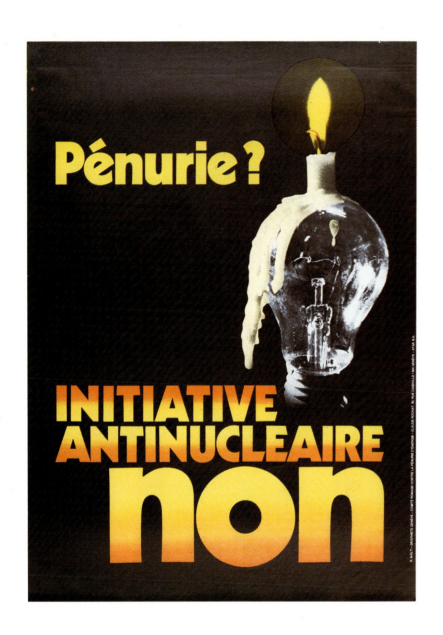

67.

R. Wälti
1979, 90,5 × 128 cm
Offset printing / Offsetdruck / Impression offset

68.

Anonym
1979, 90,5 × 128 cm
Silkprint / Siebdruck / Sérigraphie

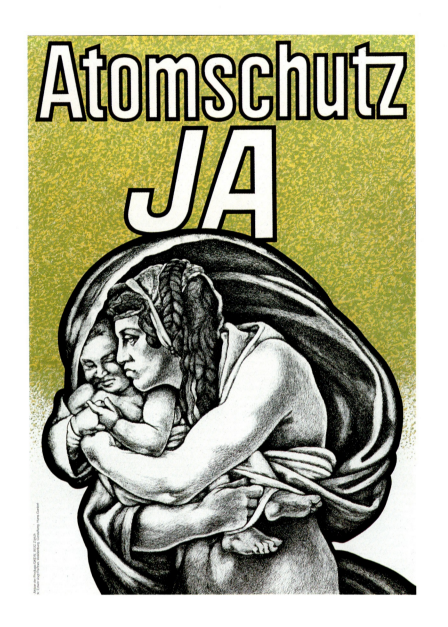

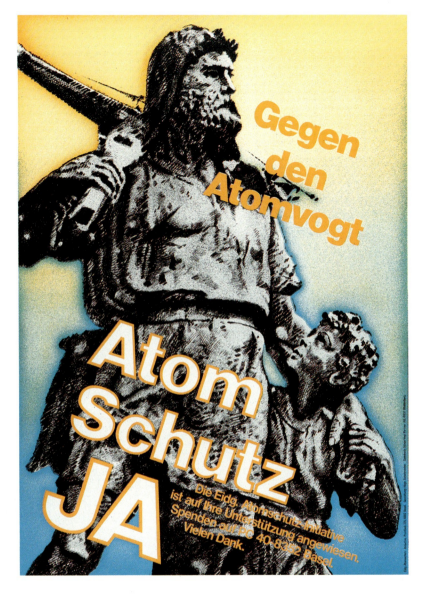

69.

Hans Gantert
1979, 90,5 × 128 cm
Silkprint / Siebdruck / Sérigraphie

70.

Hugo Schuhmacher
1979, 90,5 × 128 cm
Silkprint / Siebdruck / Sérigraphie

71.

Bernard Schlup
1979, 42×60 cm
Offset printing / Offsetdruck / Impression offset

72.

Erika Sutter
1979, 44×63 cm
Offset printing / Offsetdruck / Impression offset

73.

Jürg Stauffer
1979, 42×60 cm
Offset printing / Offsetdruck / Impression offset

74.

Pierre Brauchli
1979, 42×60 cm
Silkprint / Siebdruck / Sérigraphie

75.

Anonym
1979, 42×59 cm
Offset printing / Offsetdruck / Impression offset

76.

Peter König
1979, 42×60 cm
Silkprint / Siebdruck / Sérigraphie

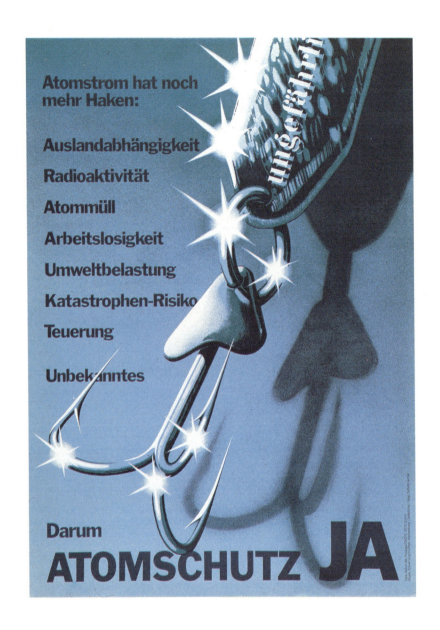

77.

Hugo Schuhmacher
1979, 90,5 × 128 cm
Silkprint / Siebdruck / Sérigraphie

78.

Ruedi Baumann
1979, 90,5 × 128 cm
Silkprint / Siebdruck / Sérigraphie

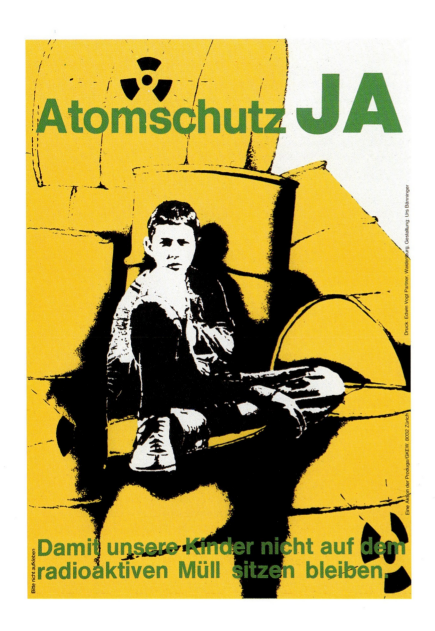
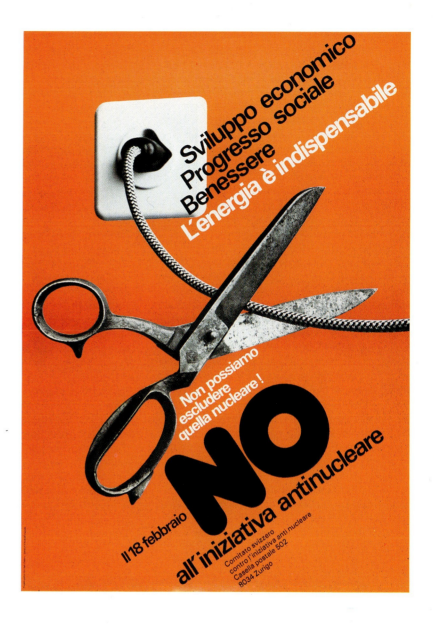

79.

Urs Bänninger
1979, 42×60 cm
Silkprint / Siebdruck / Sérigraphie

80.

Orio Galli
1981, 90,5×128 cm
Offset printing / Offsetdruck / Impression offset

First of May

Since 1920, posters have been produced fairly regularly for First of May festivities at both a local and a national level. Their function has been and is to remind workers of their worldwide festival and to summon them to active participation. At the same time, the First of May posters are intended to draw the attention of the whole population, over and above the left-wing parties and the trade unions, to this festival of the workers. Thus these posters have less of an advertising function than the character of a statement of intent: We are here, and we are strong. We must be reckoned with.
The central motif of these posters is that of the red flags which give colour to the streets during the First of May processions. In some cases this is complemented by a visualization of the main theme of the May festival of the given year. A strong optical effect is essential and there is generally no use of detailed text messages. It is unmistakable that the graphic style of these May posters is based on a specific tradition, that of the poster appeals made in the early 1920s by constructivist artists on behalf of the festivities of the young Soviet Union. In those cases, as a rule, a black-and-white photo-montage was merely strengthened, in its optical effect, by an accentuation of red.
The Swiss May posters are not the work of professional advertising artists. Mostly they are designed by free artists who are associated with socialism and testify to their political engagement through poster work. Several Zurich artists have created especially effective posters, among them being Hans Gantert, Gottfried Honegger and Hugo Schuhmacher. The most lapidary and effective of these artists' posters are by the painter Carlos Duss, known to the art-lovers of his city as a master of delicate colouring.

Der 1. Mai

Seit 1920 sind fast regelmässig für die lokalen Maifeiern in den grösseren Städten oder gesamtschweizerisch Plakate hergestellt worden. Ihre Aufgabe bestand und besteht darin, die Werktätigen an ihren weltweit gefeierten Festtag zu erinnern und zur aktiven Beteiligung aufzufordern. Gleichzeitig sollen die Plakate zum 1. Mai über die veranstaltenden Linksparteien und Gewerkschaften hinaus die gesamte Bevölkerung auf diesen Feiertag der Werktätigen aufmerksam machen. So haben diese Plakate weniger eine Werbefunktion als vielmehr den Charakter einer Positionsmeldung: Wir sind da, und wir sind stark. Es ist mit uns zu rechnen.
Zentrales Motiv dieser Plakate ist die bei den Maiumzügen das Strassenbild prägende rote Fahne. Unter Umständen kommt dazu eine Visualisierung des Leitgedankens, unter den die Maifeiern eines Jahres jeweils gestellt sind. Wesentlich ist die starke optische Wirkung; auf ausführliche Textbotschaften wird in der Regel verzichtet. Es ist unverkennbar, dass der graphische Stil dieser Maiplakate auf einer bestimmten Tradition fusst: auf den Plakataufrufen, die in den frühen zwanziger Jahren in Russland von den konstruktivistischen Künstlern für Feiern der jungen Sowjetunion geschaffen wurden. Dort wurde in der Regel eine schwarzweisse Photo- und Schriftmontage lediglich mit einem Rotakzent in der optischen Wirkung gesteigert.
Die schweizerischen Maiplakate sind nicht Produkte professioneller Werbegraphiker. Meist sind es freie Künstler, die sich zum Sozialismus bekennen und ihr politisches Engagement durch Plakatentwürfe bekunden. Besonders wirkungsvolle Plakate schufen ein paar Zürcher Maler: Hans Gantert, Gottfried Honegger, Hugo Schuhmacher und andere. Das lapidarste, wirkungsvollste unter diesen Künstlerplakaten stammt von dem Maler Carlos Duss, den die Zürcher Kunstfreunde als einen Meister delikater Farbsetzung kennen.

Le 1er mai

Depuis 1920, des affiches locales, régionales et même nationales ont été créées assez régulièrement à l'occasion des fêtes du 1er mai. Leur rôle consiste à rappeler aux travailleurs ce jour de fête commémoré dans le monde entier et à les inviter à y participer activement. En même temps, par-delà les partis politiques de gauche et les syndicats qui organisent les manifestations du 1er mai, elles visent à attirer l'attention de l'ensemble de la population sur ce jour de fête des travailleurs. Loin d'assumer simplement un rôle de propagande, ces affiches signalent une prise de position: Nous sommes là, nous sommes forts. Il faut compter avec nous.
Le motif central de ces affiches est le drapeau rouge, arboré pendant les défilés à travers les rues. Parfois, une visualisation du slogan spécifique de la Fête du travail de l'année en cours vient s'y associer. L'effet optique est déterminant; les messages sous forme de textes détaillés sont en général évités. Il est incontestable que le style graphique des affiches pour le 1er mai se fonde sur une tradition bien établie: c'est le style créé en U.R.S.S. dans les années vingt par les artistes constructivistes pour les festivités de la jeune Union soviétique. Les appels lancés à l'époque par voie d'affiches consistaient en règle générale en un photomontage noir/blanc, incluant le texte et renforcé dans son effet optique par un élément marquant rouge.
Les affiches suisses pour la Fête du travail ne sont pas le produit d'affichistes professionnels; le plus souvent, elles sont l'œuvre d'artistes indépendants qui professent le socialisme et expriment leur engagement politique par la création d'affiches. Des affiches particulièrement éloquentes ont été réalisées par les peintres zurichois Hans Gantert, Gottfried Honegger, Hugo Schuhmacher et d'autres. L'affiche la plus percutante et la plus lapidaire est l'œuvre du peintre Carlos Duss, bien connu dans les milieux artistiques zurichois pour sa subtile maîtrise des couleurs.

81.

Carl Scherer
1920, 90,5 × 128 cm
Lithography / Lithographie / Lithographie

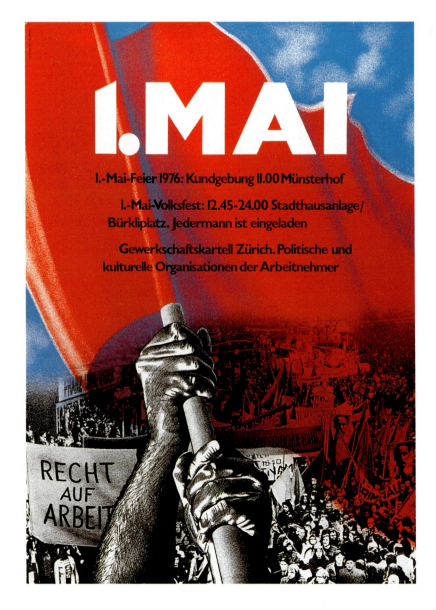

82.

Carl Scherer
1939, 90,5 × 128 cm
Lithography / Lithographie / Lithographie

83.

Hugo Schuhmacher
1976, 90,5 × 128 cm
Silkprint / Siebdruck / Sérigraphie

84.

Carlos Duss
1977, 90,5 × 128 cm
Silkprint / Siebdruck / Sérigraphie

85.

Hans Gantert
1980, 90,5 × 128 cm
Offset printing / Offsetdruck / Impression offset

86.

Gottfried Honegger
1981, 90,5 × 128 cm
Offset printing / Offsetdruck / Impression offset

87.

E. Gerber
1982, 42 × 59 cm
Offset printing / Offsetdruck / Impression offset

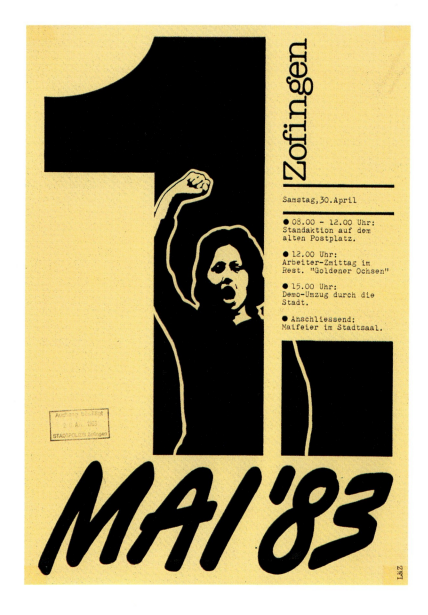

88.

Jakob Lämmler
1982, 36×56 cm
Offset printing / Offsetdruck / Impression offset

89.

Signatur Läz
1983, 30×42 cm
Offset printing / Offsetdruck / Impression offset

Winter Aid

Winter Aid is one of the oldest-established and most deserving of Swiss social and charitable organizations. It was founded in the middle of the economic crisis of the 1930s, in the particularly difficult period 1935/36, by a committee of initiative (23rd October, 1936). At first conceived as a "one-off" aid programme, Winter Aid became an institution as soon as 1937. Such it has remained to our own time, undergoing numerous revisions of aim and principal activity during the changes of the intervening years. The underlying philosophy is as follows: "Private aid is intended to fill in the gaps of government welfare. The state must generalize but the private institute must individualize." Both the raising of funds and the provision of aid are based on the voluntary principle.

The public collections of Winter Aid have almost always been accompanied by poster campaigns. Ever since the impressive visualization of the slogan "Help us to help" ("Helft uns helfen") in 1936 (Behrmann/Bosshard), the collections have generally been supported by posters of superior artistic quality, mostly designed by leading graphic artists from all parts of the country. After the earlier use of pictorial themes, such as the woman waiting in the winter cold for a bowl of soup (Franz Fässler), symbolic motifs come to the fore in later works, for example the candle as a symbol of security (Gérard Miedinger, Emil Hotz). Artistic arrangements of type matter have been repeatedly used (Ernst Keller, Andreas His). Occasionally the snow crystal has also been used as a graphically impressive symbol (Paul Sollberger).

A noticeable revival was achieved when Winter Aid began to obtain poster designs through competitions at schools of arts and crafts. This has led to the production, since the middle 1950s, of a wide variety of posters, mostly powerful in form and feeling, with which Winter Aid has repeatedly reached the first rank in Swiss poster art. It is a matter for special attention that an institution devoted to welfare in the widest fields gives its publicity such a brave face, at the same time offering young designers their first chances of productive work.

Die Winterhilfe

Unter den Schweizer Hilfs- und Sozialwerken ist die Winterhilfe eines der ältesten und verdienstvollsten. Mitten in der Wirtschaftskrise der dreissiger Jahre, im besonders schweren Krisenjahr 1935/36, wurde sie am 23. Oktober 1936 von einem Initiativkomitee gegründet. Aus einem zuerst als einmalige Aktion gedachten Unternehmen wurde schon 1937 eine feste Institution. Sie ist es mit immer wieder überprüften Zielsetzungen und Schwerpunkten in stets wieder anderen Zeiten bis heute geblieben. Grundgedanke ist: «Die private Fürsorge soll Lücken staatlicher Wohlfahrtspflege schliessen. Der Staat muss verallgemeinern, die private Fürsorge muss individualisieren.» Sowohl die Mittelbeschaffung wie die Hilfseinsätze beruhen auf dem Prinzip der Freiwilligkeit.

Die öffentlichen Sammelaktionen der Winterhilfe waren fast immer begleitet von Plakataktionen. Seit der einprägsamen Visualisierung des Slogans «Helft uns helfen» von 1936 (Behrmann/Bosshard) begleiten in der Regel gestalterisch vorzügliche Plakate die Sammelaktionen. Beteiligt waren meist führende Graphiker aus allen Landesteilen. Nachdem anfänglich ein Bildmotiv – etwa die in der Winterkälte auf die Suppenabgabe wartende Frau (Franz Fässler) – verwendet wurde, traten später symbolische Motive in den Vordergrund, etwa die Kerze als Symbol der Geborgenheit (Gérard Miedinger, Emil Hotz). Immer wieder wurden künstlerische Schriftgestaltungen versucht (Ernst Keller, Andreas His). Zeitweise wurde auch der Schneestern zu graphisch eindrücklichen Symbolzeichen (Paul Sollberger).

Eine sichtliche Verlebendigung ergab sich, als die Winterhilfe dazu überging, die Plakatentwürfe durch Wettbewerbe an den Kunstgewerbeschulen zu gewinnen. So entstanden seit Mitte der fünfziger Jahre sehr unterschiedliche, formal und stimmungsmässig meist starke Plakate, mit denen die Winterhilfe immer wieder an vorderster Stelle der Schweizer Plakatproduktion stand. Dass gerade eine Institution, die aufs Wohlwollen breitester Kreise angewiesen ist, ihrer Werbung ein so mutiges Gesicht gibt (und gleichzeitig jungen Gestaltern erste Realisierungsmöglichkeiten bietet), das verdient besondere Beachtung.

Le Secours suisse d'hiver

Parmi les œuvres sociales d'assistance et de secours en Suisse, le Secours suisse d'hiver est l'une des plus anciennes et des plus méritoires. Fondée par un comité d'initiative le 23 octobre 1936, c'est-à-dire en pleine crise économique des années trente, l'entreprise était d'abord conçue comme action unique, puis devint une institution fixe à partir de 1937. Depuis lors, sa conception fondamentale est restée la même à travers des périodes souvent changeantes, mais avec des objectifs et points forts constamment mis à jour: «La prévoyance privée doit combler les lacunes de l'assistance publique. L'Etat doit généraliser, la prévoyance privée doit individualiser.» Tant la collecte de fonds que les actions de secours se fondent sur le principe de l'aide bénévole.

Les campagnes de collecte publiques du Secours suisse d'hiver étaient presque toujours appuyées par des affiches. Depuis la visualisation incisive du slogan «Aidez-nous à aider» de 1936 (Behrmann/Bosshard), les campagnes de collecte de fonds sont en général accompagnées d'affiches d'excellente conception formelle. Des graphistes notoires de toutes les régions du pays y apportent leur contribution. Alors qu'au début, l'illustration constituait le motif central – comme par exemple l'image d'une femme attendant la distribution de la soupe dans le froid hivernal (Franz Fässler) – les motifs symboliques prédominent plus tard, comme la bougie en tant que symbole de sécurité et d'intimité (Gérard Miedinger, Emil Hotz). A diverses reprises, les essais ont également porté sur la conception typographique (Ernst Keller, Andreas His). Parfois, le cristal de neige a servi de symbole graphique expressif (Paul Sollberger).

La mise au concours des projets d'affiche dans les écoles des arts et métiers a clairement animé la scène. A partir du milieu des années cinquante, des affiches très diverses, de conception formelle et émotionnelle percutante, ont permis au Secours suisse d'hiver de figurer aux premiers rangs pour la production d'affiches suisses. Qu'une institution qui dépend de la bienveillance de larges couches de la population, réussisse à conférer à sa publicité un profil aussi audacieux, tout en offrant à de jeunes créateurs leurs premières chances de réalisation graphique, voilà qui mérite d'être spécialement relevé.

90.

Behrmann/Bosshard
1936, 90,5 × 128 cm
Lithography / Lithographie / Lithographie

 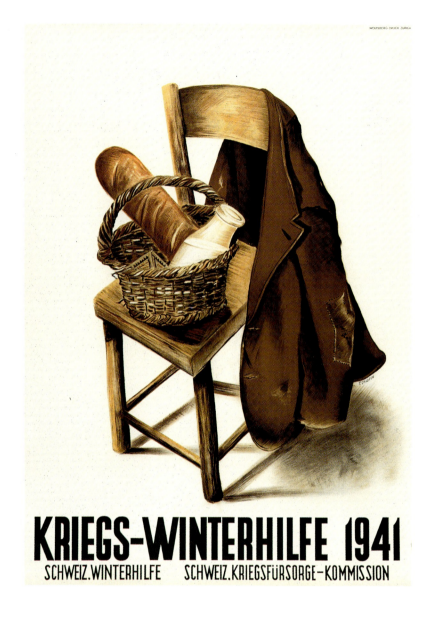

91.

Roland Guignard
1937, 90,5 × 128 cm
Lino cut / Linolschnitt / Gravure sur linoléum

92.

Paul Jacopin
1941, 90,5 × 128 cm
Lithography / Lithographie / Lithographie

93.

Franz Fässler
1942, 90,5 × 128 cm
Lithography / Lithographie / Lithographie

94.

Pierre Monnerat
1943, 90,5 × 128 cm
Lithography / Lithographie / Lithographie

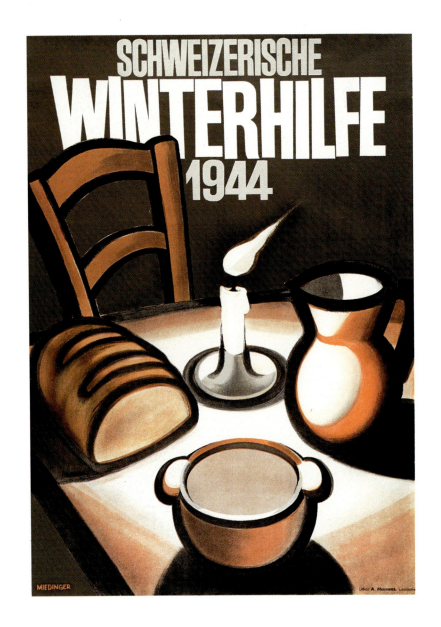

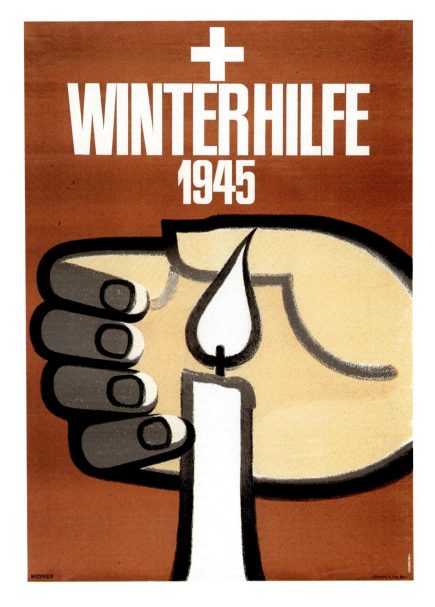

95.

Gérard Miedinger
1944, 90,5 × 128 cm
Lithography / Lithographie / Lithographie

96.

Gérard Miedinger
1945, 90,5 × 128 cm
Lithography / Lithographie / Lithographie

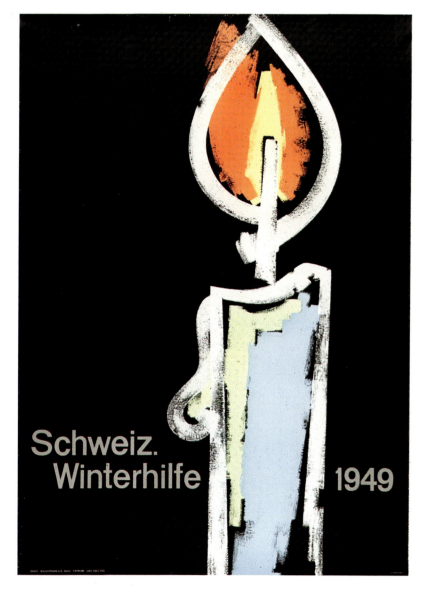

97.

Ernst Keller
1946, 90,5 × 128 cm
Lino cut / Linolschnitt / Gravure sur linoléum

98.

Emil Hotz
1949, 90,5 × 128 cm
Lithography / Lithographie / Lithographie

99.

Paul Sollberger
1950, 90,5 × 128 cm
Lithography / Lithographie / Lithographie

100.

Paul Sollberger
1951, 90,5 × 128 cm
Lithography / Lithographie / Lithographie

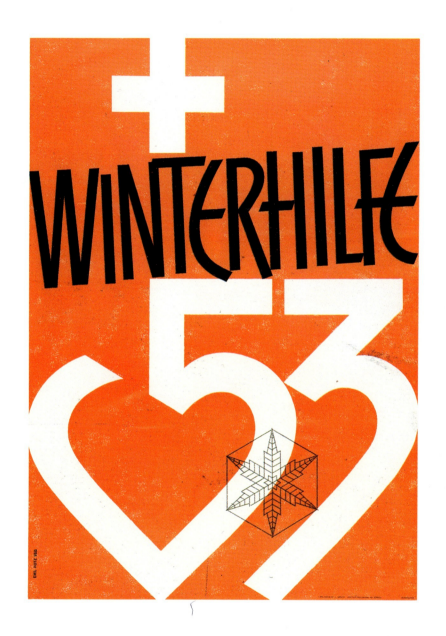

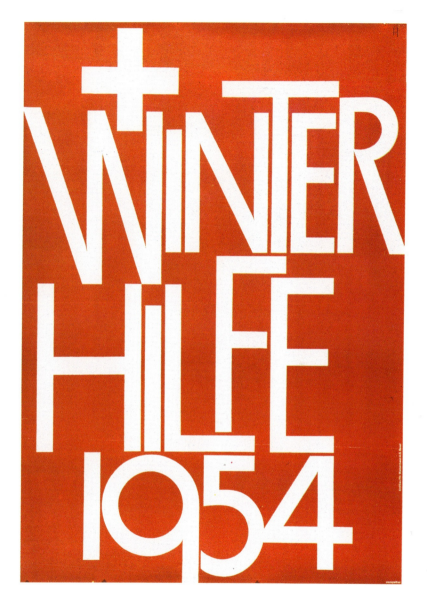

101.

Emil Hotz
1953, 90,5 × 128 cm
Lino cut / Linolschnitt / Gravure sur linoléum

102.

Andreas His
1954, 90,5 × 128 cm
Offset printing / Offsetdruck / Impression offset

103.

Wehrli
1956, 90,5 × 128 cm
Lithography / Lithographie / Lithographie

104.

Peter Andermatt
1958, 90,5 × 128 cm
Intaglio / Tiefdruck / Impression en creux

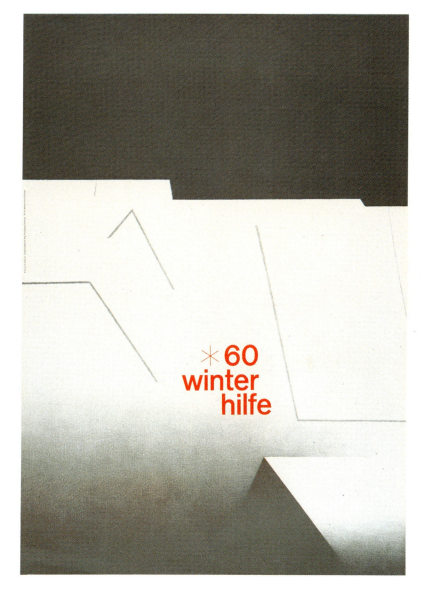

105.

Peter Andermatt
1959, 90,5 × 128 cm
Offset printing / Offsetdruck / Impression offset

106.

Ulrich Schierle
1960, 90,5 × 128 cm
Offset printing / Offsetdruck / Impression offset

107.

Peter Hajnoczky
1962, 90,5 × 128 cm
Lithography / Lithographie / Lithographie

108.

Andreas Cathomas
1964, 90,5 × 128 cm
Offset printing / Offsetdruck / Impression offset

109.

Emil Steinberger / Niklaus Birrer
1965, 90,5 × 128 cm
Offset printing / Offsetdruck / Impression offset

110.

Ruth Pfalzberger
1969, 90,5 × 128 cm
Offset printing / Offsetdruck / Impression offset

111.

Steffen Wolff
1971, 90,5 × 128 cm
Offset printing / Offsetdruck / Impression offset

112.

Walter Wild
1972, 90,5 × 128 cm
Offset printing / Offsetdruck / Impression offset

113.

Heinz Bäder
1979, 90,5 × 128 cm
Offset printing / Offsetdruck / Impression offset

114.

Basil Hangarter
1980, 90,5 × 128 cm
Offset printing / Offsetdruck / Impression offset

For the Old

Socially engaged people were aware, as long ago as the difficult years immediately after the First World War, that something must be done for the old, when they are no longer able to earn their living. Collections on behalf of the private Foundation for the Elderly came to supplement the largely cantonal support funds for old people, in so far as they existed at all, in 1921.
Publicity for these collections has always included posters. Since 1921, painters such as Hans Beat Wieland and graphic artists like Pierre Gauchat, Pierre Kramer, Alfred Widmer and others, have made their services available. The subject at first suggested an evocative, representative treatment of the theme of old age. Feelings of respect and sympathy, and perhaps also of a guilty conscience, were intended to be awakened. It was only slowly that an artistic claim was added to that of the content: in the 1940s there are attempts to handle the theme of old age in a primarily artistic way, without sentimentality. Otto Baumberger's poster of 1941 is an example of this tendency.
Bolder solutions began in 1945, with the unforgettable poster by Hans Falk, which in its graphics shows a feeling rather than a subject. From then on, the chain of convincing posters remains unbroken, whether they use photographic means (as with Carlo Vivarelli and Celestino Piatti), or graphics (Hans Falk) or in some cases (Ernst Keller and Hans Aeschbach, for example) the resources of painting.
With the extension of national insurance against old age and bereavement, the public activities of the Foundation for the Elderly lose their former importance, and it is no coincidence that there has been a decline in its posters, which had long been among the most important contributions to Swiss poster art, since the 1960s. Among the last of these powerful creations is Fridolin Müller's photo-poster of old and tired hands holding a pair of spectacles.

Für das Alter

Dass gerade für den alten, nicht mehr im Erwerbsleben stehenden Menschen etwas getan werden müsste, wurde sozial engagierten Persönlichkeiten schon in den schwierigen Jahren unmittelbar nach dem Ersten Weltkrieg bewusst. Den weitgehend kantonalen Unterstützungskassen für die Alten, soweit es sie überhaupt schon gab, traten seit 1921 die Sammelaktionen der privaten Stiftung für das Alter zur Seite.
Die Werbung für diese Sammelaktionen wurde stets begleitet von Plakaten. Seit 1921 stellten sich dafür Maler (wie Hans Beat Wieland) und Graphiker-Maler (wie Pierre Gauchat, Pierre Kramer, Alfred Widmer u. a.) zur Verfügung. Die Sache legte zunächst eine suggestive Bilddarstellung des alten Menschen nahe. Gefühle des Respekts, des Mitleids und vielleicht des schlechten Gewissens sollten ausgelöst werden. Nur langsam trat zu diesem inhaltlichen Anspruch auch ein künstlerischer: In den vierziger Jahren tauchen Versuche auf, das Thema alter Mensch primär künstlerisch und nicht sentimental zu bewältigen: Das Plakat von Otto Baumberger von 1941 ist ein solcher Versuch.
Kühnere Lösungen setzen 1945 mit dem unvergesslichen, betont zeichnerischen Plakat von Hans Falk ein, das weniger eine Sache als eine Stimmung zeigt. Von da an reisst die Kette überzeugender Plakate nicht mehr ab, seien sie (wie bei Carlo Vivarelli und Celestino Piatti) mit photographischen Mitteln oder (wie bei Hans Falk) mit zeichnerischen, teils auch (wie bei Ernst Keller und Hans Aeschbach) mit malerischen Mitteln gelöst.
Mit dem Ausbau der staatlichen Alters- und Hinterbliebenen-Versicherung verlieren die öffentlichen Aktivitäten der Stiftung für das Alter an Bedeutung. Kein Zufall, dass seit den sechziger Jahren auch die Plakate, die lange zu den wesentlichen Beiträgen der Schweizer Plakatproduktion gehört hatten, aussetzen. Zu den letzten starken Lösungen gehört das Photoplakat von Fridolin Müller mit den altersmüden, eine Brille haltenden Händen.

Pour la vieillesse

Dans les années difficiles après la Première Guerre mondiale, la nécessité d'intervenir en faveur de ceux qui se sont retirés de la vie professionnelle active avait déjà été ressentie par divers protagonistes très engagés sur le plan social. L'action des caisses d'assistance cantonales – pour autant que celles-ci existaient déjà – était complétée par des campagnes de collecte de fonds de la Fondation nationale suisse pour la vieillesse.
La propagande pour les collectes de cette œuvre privée a toujours été appuyée par voie d'affiches. Depuis 1921, des peintres (comme Hans Beat Wieland) et des graphistes-peintres (comme Pierre Gauchat, Pierre Kramer, Alfred Widmer et d'autres) se sont mis au service de la cause de l'homme âgé. Par des images suggestives, ils s'employaient d'abord à susciter des sentiments de respect, de pitié, peut-être même de mauvaise conscience. Peu à peu seulement, dans les années quarante, les affichistes ont commencé à donner au thème de l'homme âgé une interprétation non plus sentimentale, mais artistique: l'affiche d'Otto Baumberger exprime une telle tentative.
Des solutions plus audacieuses interviennent en 1945 avec l'inoubliable affiche de Hans Falk, évoquant par le dessin une ambiance beaucoup plus qu'une cause. C'était le début d'une suite ininterrompue d'affiches convaincantes, réalisées par les moyens de la photographie (Carlo Vivarelli et Celestino Piatti), du dessin (Hans Falk) ou de la peinture (Ernst Keller et Hans Aeschbach).
Avec l'institution de l'Assurance vieillesse-survivants (AVS) par l'Etat, les activités publiques de la Fondation pour la vieillesse perdent en importance. Aussi ne faut-il guère s'étonner d'assister, à partir des années soixante, à un arrêt de la production d'affiches pour cette œuvre. Une des dernières œuvres marquantes, l'affiche photographique de Fridolin Müller, montre les mains ridées d'un vieillard tenant une paire de lunettes.

115.

Hans Beat Wieland
1921, 90,5 × 128 cm
Lithography / Lithographie / Lithographie

116.

Jules Courvoisier
1925, 90,5 × 128 cm
Lithography / Lithographie / Lithographie

117.

Pierre Gauchat
1925, 90,5 × 128 cm
Draft of poster / Plakatentwurf / Projet d'affiche

118.

E. Weiss
1926, 87 × 128 cm
Lithography / Lithographie / Lithographie

119.

Hans Beat Wieland
1927, 90,5 × 128 cm
Lithography / Lithographie / Lithographie

120.

Pierre Kramer
1931, 90,5 × 128 cm
Lithography / Lithographie / Lithographie

121.

Alfred Widmer
1932, 90,5 × 128 cm
Lithography / Lithographie / Lithographie

122.

Alfred Widmer
1934, 90,5 × 128 cm
Lithography / Lithographie / Lithographie

123.

Pierre Kramer
1935, 90,5 × 128 cm
Lithography / Lithographie / Lithographie

124.

Ernst Emil Schlatter
1939, 90,5 × 128 cm
Lithography / Lithographie / Lithographie

125.

Otto Baumberger
1941, 90,5 × 128 cm
Lithography / Lithographie / Lithographie

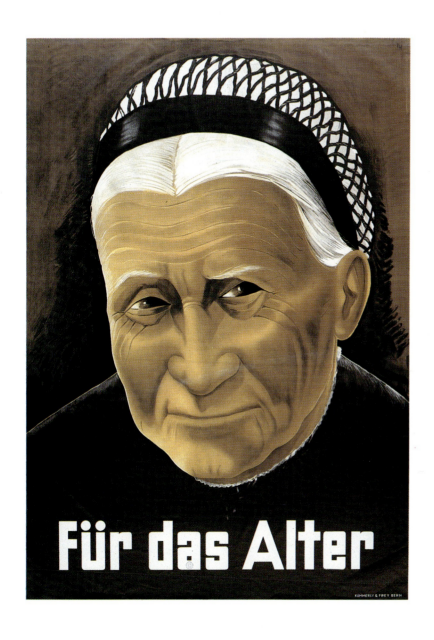
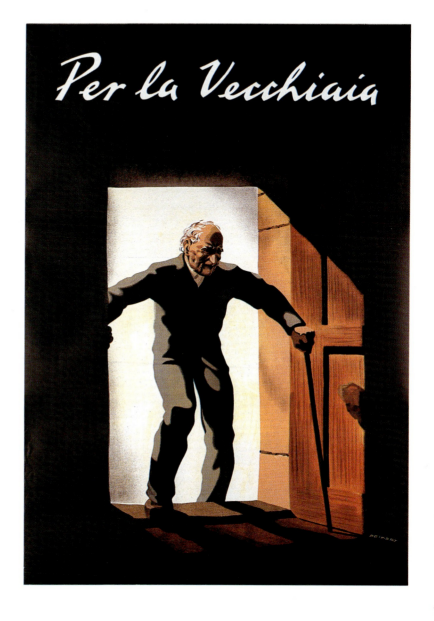

126.

Hans Handschin
1942, 90,5 × 128 cm
Lithography / Lithographie / Lithographie

127.

Martin Peikert
1943, 90,5 × 128 cm
Lithography / Lithographie / Lithographie

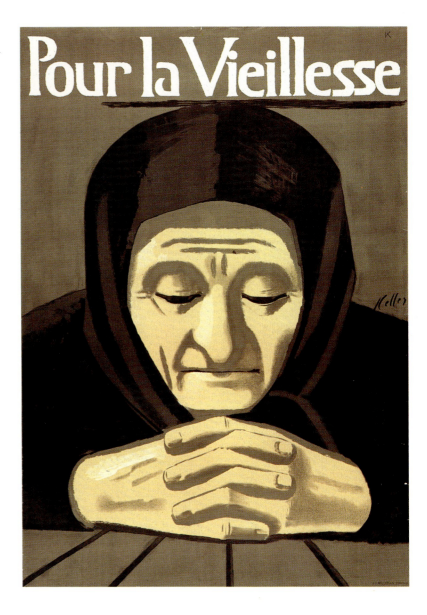

128.

Hans Falk
1945, 90,5 × 128 cm
Lithography / Lithographie / Lithographie

129.

Ernst Keller
1946, 90,5 × 128 cm
Lithography / Lithographie / Lithographie

130.

Hans Aeschbach
1947, 90,5 × 128 cm
Lithography / Lithographie / Lithographie

131.

Hans Falk
1948, 90,5 × 128 cm
Lithography / Lithographie / Lithographie

132.

Carlo Vivarelli
1949, 90,5 × 128 cm
Intaglio / Tiefdruck / Impression en creux

133.

Celestino Piatti / Photo: Stricker
1953, 90,5 × 128 cm
Offset printing / Offsetdruck / Impression offset

134.

Ruedi Barth
1954, 90,5×128 cm
Lithography / Lithographie / Lithographie

135.

Fridolin Müller
1964, 90,5×128 cm
Offset printing / Offsetdruck / Impression offset

Aid for Refugees

The official Swiss policy on refugees during the war years is still a subject of dispute, perhaps more today than ever before. Independently of the behaviour of the national government, however, and its association with alleged "reasons of state", many private organizations, often formed on an *ad hoc* basis, have tried to alleviate the lot of refugees and homeless people from a variety of places. To give publicity to collections of money, impressive posters have frequently been designed by leading graphic artists, mostly not as commissions but as free contributions to the good cause.

Among these effective and artistically valid posters are the expressive designs of Gérard Miedinger and, particularly, the almost monochrome, graphically striking works of Hans Falk. The poster created in 1946 for a collection on behalf of the homeless has become known as a prototype of the social poster, well beyond its time and country, comparable in its density of meaning to the social posters of Théophile Steinlen from the turn of the century.

Hilfe für Flüchtlinge

Die offizielle Flüchtlingspolitik der Schweiz in den Kriegsjahren ist noch immer umstritten, ja ist es vielleicht heute mehr denn je. Unabhängig von der durch angebliche «Staatsraison» bestimmten Haltung der Landesregierung haben jedoch zahlreiche private, oft ad hoc gebildete Organisationen das Los von Flüchtlingen und Heimatlosen unterschiedlicher Herkunft zu lindern versucht. Für Geldsammlungen entstanden häufig einprägsame Plakate führender graphischer Gestalter – meist nicht Auftragsarbeiten, sondern vielmehr unentgeltlich der guten Sache zur Verfügung gestellte Entwürfe. Zu diesen wirkungsvollen und zugleich plakatkünstlerisch gültigen Plakaten gehören die expressiven Entwürfe von Gérard Miedinger und vor allem die fast unfarbigen, zeichnerisch ausdrucksstarken Blätter von Hans Falk. Das 1946 für eine Sammelaktion zugunsten der Heimatlosen geschaffene Plakat ist, weit über die Schweiz und die Zeit hinaus, zu einem Prototyp des sozialen Plakats geworden – vergleichbar in seiner Dichte den sozialen Plakaten von Théophile Steinlen aus dem Jahrhundertanfang.

L'aide aux réfugiés

La politique officielle suisse d'aide aux réfugiés pendant les années de guerre a été souvent critiquée, et l'est peut-être aujourd'hui plus que jamais. Indépendamment de l'attitude du gouvernement fédéral, déterminée par la prétendue «raison d'Etat», de nombreuses organisations privées, souvent constituées ad hoc, ont cherché à alléger le sort des réfugiés et apatrides venus des pays les plus divers. Des affiches suggestives de collecte de fonds ont été créées par des graphistes notoires qui, sans avoir été spécialement mandatés, se sont mis spontanément et souvent gracieusement à la disposition de la bonne cause.

Parmi ces affiches politiques à la fois percutantes et de haute qualité artistique, il convient de citer les travaux expressifs de Gérard Miedinger et surtout les réalisations de Hans Falk, qui se distinguent par la force d'expression du dessin et la sobriété des couleurs. L'affiche conçue par Hans Falk en 1946 pour une collecte de fonds en faveur des apatrides est devenue, bien au-delà des frontières suisses et de son époque, le prototype de l'affiche sociale – comparable dans sa densité aux affiches sociales de Théophile Steinlen au début de ce siècle.

136.

Gérard Miedinger
1942, 90,5 × 128 cm
Lithography / Lithographie / Lithographie

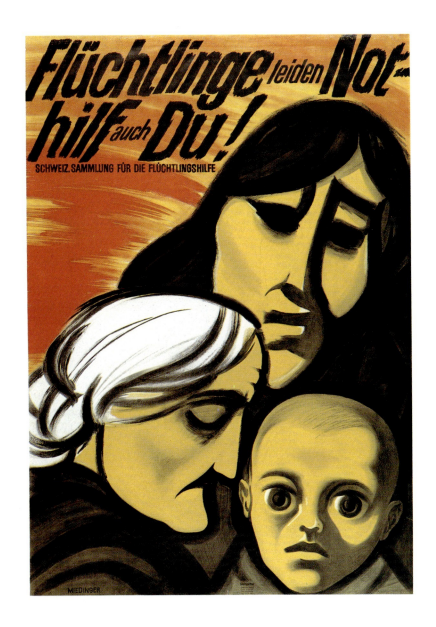

137.

Gérard Miedinger
1943, 90,5 × 128 cm
Lithography / Lithographie / Lithographie

138.

Hans Falk
1944, 90,5 × 128 cm
Lithography / Lithographie / Lithographie

139.

Hans Falk
1945, 90,5 × 128 cm
Lithography / Lithographie / Lithographie

140.

Hans Falk
1946, 90,5 × 128 cm
Lithography / Lithographie / Lithographie

The Red Cross

In its humanitarian concept, the International Committee of the Red Cross, founded in 1864 by a group of Swiss inspired by the idea of Henri Dunant, is renowned all over the world as one of the greatest achievements of Switzerland. Both the International Red Cross and the Swiss national branch of the organization are repeatedly publicized in Switzerland in connection with various kinds of appeal, such as the call for blood donors. Most of the posters have supported appeals for money, goods or blood donations.
A campaign conducted in 1943 gave Otto Baumberger the opportunity to display the various fields of activity of the ICRC in expressively illustrative posters. In the succeeding years, Hans Neuburg, Ernst Keller and a few others created posters which were certainly more significant from an artistic viewpoint, but after that period, a remarkable gap occurred in the poster production of the Red Cross.
After nearly two decades of inactivity, the very unconventional, lively posters of André Masmejan, Michel Gallay, Edgar Küng and others, some using photographic methods, appeared in the Sixties. Many friends of poster art have noted with admiring respect that a time-honoured institution has become associated with the language of modern graphics through such original and effective posters.

Das Rote Kreuz

Das 1864, angeregt durch Henri Dunants Idee, durch eine Gruppe von Schweizern geschaffene Internationale Komitee vom Roten Kreuz gilt in seinem humanitären Konzept weltweit als eine der grossen Leistungen der Schweiz. Sowohl das Internationale Rote Kreuz wie das innerhalb der Schweiz tätige Schweizerische Rote Kreuz sind immer wieder mit Aktionen unterschiedlichen Charakters – etwa den Blutspende-Aktionen – an die Schweizer Öffentlichkeit gelangt. Plakate haben meist die Bitten um Geld, Hilfsgüter oder Blut unterstützt.
Eine im Jahre 1943 durchgeführte Plakataktion gab Otto Baumberger die Möglichkeit, die verschiedenen Tätigkeitsbereiche des IKRK in expressiv-illustrativen Plakaten zu veranschaulichen. Plakatkünstlerisch wohl bedeutendere Entwürfe schufen in den folgenden Jahren Hans Neuburg, Ernst Keller und wenige andere. Dann war es merkwürdig still um die Plakatproduktion des Roten Kreuzes. Nach einem Unterbruch von beinahe zwei Jahrzehnten entstanden jedoch in den sechziger Jahren sehr unkonventionelle, lebendige, teils mit photographischen Mitteln arbeitende Plakate von André Masmejan, Michel Gallay, Edgar Küng und andern. Dass mit solchen frischen und zugleich wirkungsvollen Plakaten eine altehrwürdige Institution sich zur graphischen Sprache von heute bekannt hat, das ist mit bewunderndem Respekt von vielen Freunden des Plakats zur Kenntnis genommen worden.

La Croix-Rouge

Le Comité international de la Croix-Rouge, dont la conception humanitaire compte dans le monde entier parmi les plus grandes réalisations de la Suisse, a été fondé en 1864 par un groupe de Suisses inspirés par Henri Dunant. Tant la Croix-Rouge internationale que la Croix-Rouge suisse (dont l'activité se concentre sur le territoire helvétique) n'ont cessé de solliciter l'appui de la population suisse: par des campagnes d'affiches largement diffusées, les appels les plus divers ont été lancés pour demander au public des dons sous forme d'argent, de biens ou de sang.
La campagne lancée en 1943 donna à Otto Baumberger l'occasion d'illustrer l'activité des divers secteurs du CICR par des affiches expressives. Les réalisations plus convaincantes de Hans Neuburg, Ernst Keller et de quelques autres créateurs datent des années suivantes. Puis, ce fut le grand silence. Après une interruption de près de deux décennies, la production d'affiches pour la Croix-Rouge connut un regain d'activité avec des œuvres vivantes et peu conventionnelles, réalisées en partie avec des moyens photographiques par André Masmejan, Michel Gallay, Edgar Küng et d'autres. Le fait qu'une institution d'aussi grande tradition adopte le langage graphique moderne pour ses affiches à la fois suggestives et d'une vivifiante fraîcheur a été accueilli avec respect et admiration par de nombreux amis de l'affiche.

141.

Otto Baumberger
1943, 90,5 × 128 cm
Lithography / Lithographie / Lithographie

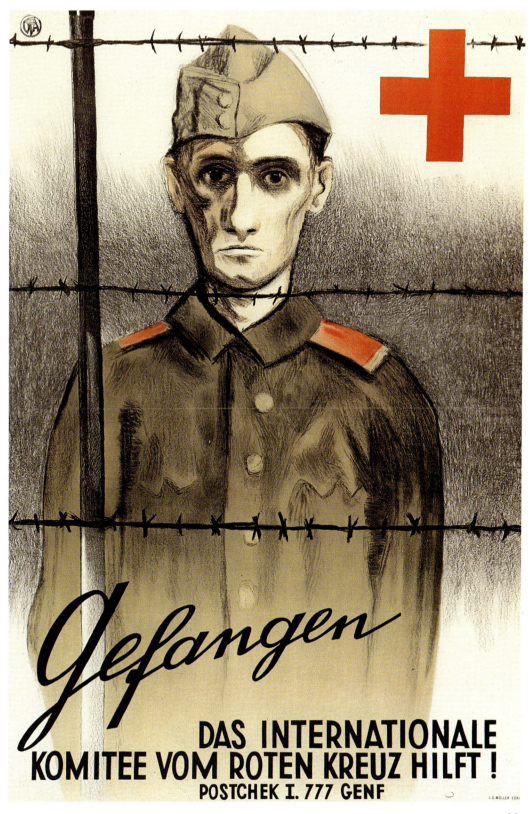

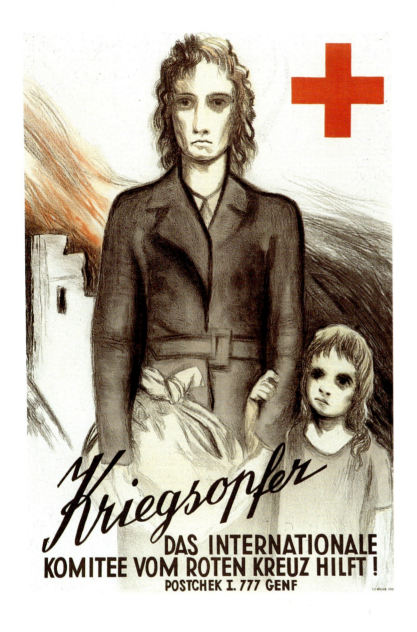

142.

Otto Baumberger
1943, 90,5 × 128 cm
Lithography / Lithographie / Lithographie

143.

Otto Baumberger
1943, 90,5 × 128 cm
Lithography / Lithographie / Lithographie

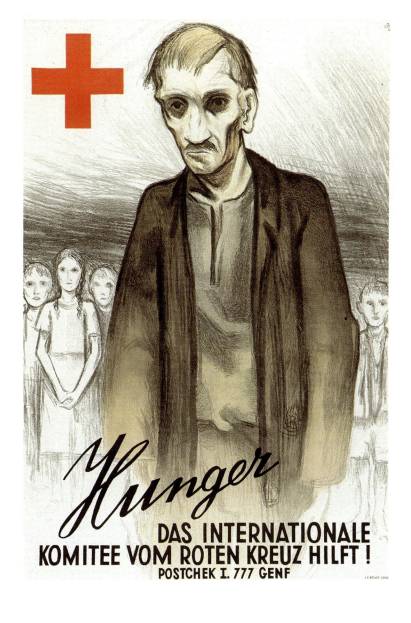

144.

Otto Baumberger
1943, 90,5 × 128 cm
Lithography / Lithographie / Lithographie

145.

Otto Baumberger
1943, 90,5 × 128 cm
Lithography / Lithographie / Lithographie

146.

Hans Neuburg
1944, 90,5 × 128 cm
Intaglio / Tiefdruck / Impression en creux

147.

Eugen Früh
1944, 90,5 × 128 cm
Lithography / Lithographie / Lithographie

148.

Ernst Keller
1946, 90,5 × 128 cm
Lithography / Lithographie / Lithographie

149.

Ernst Keller
1946, 90,5 × 128 cm
Lithography / Lithographie / Lithographie

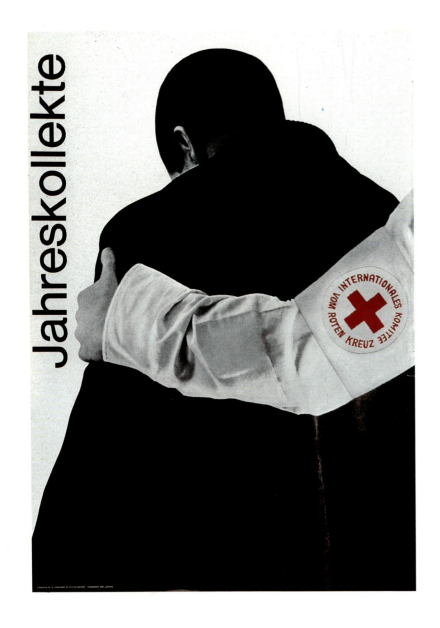

150.

André Masmejan / W. Schmid
1962, 90,5 × 128 cm
Offset printing / Offsetdruck / Impression offset

151.

Michel Gallay
1963, 90,5 × 128 cm
Offset printing / Offsetdruck / Impression offset

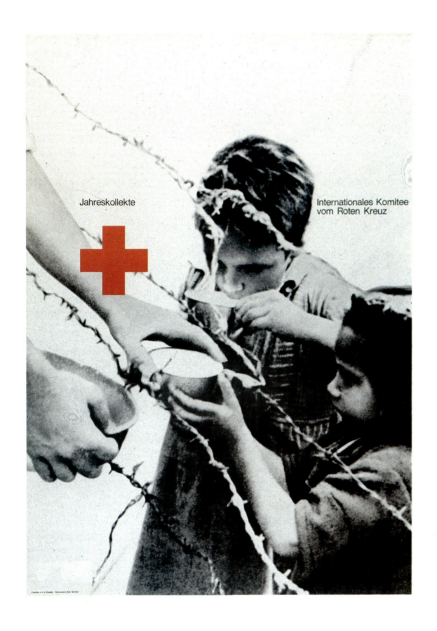

152.

André Masmejan / W. Schmid
1966, 90,5 × 128 cm
Offset printing / Offsetdruck / Impression offset

153.

Edgar Küng
1971, 90,5 × 128 cm
Silkprint / Siebdruck / Sérigraphie

Pro Infirmis

Humanitarian foundations differ from one another in other ways than in their activities and objectives and the strategies they use to achieve their chosen aims. There is also, for example, much difference in their ways of using posters as an advertising medium.
In 1940, the Pro Infirmis foundation, established since 1920, put its message before the public with a poster which, in its theme, gave its concern an authentic representation, at the same time showing great quality as a piece of poster art, with a protective hand sheltering a delicate and highly fragile blossom. This design by Alois Carigiet was soon replaced by a symbolic type of poster by Donald Brun, first issued in 1944: a wing, drawn with great care, hindered from freely unfolding by a chain. Clearly the responsible parties within the foundation considered this pictorial image to be so well suited to their concern that they limited their poster campaign in succeeding years to repeated variations of this theme, changing only the colours, particularly the background colour. Only once, in 1948, did the foundation deviate from this poster theme, with the expressive, masterfully drawn poster by Hans Falk, showing a young woman in a wheelchair, with her back to us and guarded by a dog. This poster has remained unforgettable for many people.
With an increasing number of handicapped people among our population and the growing acceptance of their integration into the life of the community, such moving posters as Hans Falk's are perhaps no longer necessary. Whether in the near or distant future other poster themes or design styles will promote the appeals of Pro Infirmis, remains to be seen.

Pro Infirmis

Humanitäre Werke unterscheiden sich nicht nur durch ihre Aufgabenstellungen und ihre Zielsetzungen, aber auch nicht nur durch die Strategien, mit denen gesteckte Ziele erreicht werden sollen. Es gibt auch grosse Unterschiede zum Beispiel in der Art der Nutzung des Plakats als Werbemittel.
1940 hat sich die bereits 1920 gegründete Stiftung Pro Infirmis mit einem Plakat der Öffentlichkeit gestellt, das – motivisch – ihr Anliegen gültig zur Darstellung brachte und zugleich als Plakat grosse künstlerische Qualitäten aufwies: die Hand, die schützend, rettend eine zierliche, höchst gebrechliche Blüte trägt. Der von Alois Carigiet stammende Entwurf ist sehr bald abgelöst worden durch ein 1944 erstmals ausgehängtes, signethaftes Plakat von Donald Brun: einen graphisch sorgfältig gestalteten Flügel, der durch eine Kette in seiner freien Entfaltung gehindert wird.
Offensichtlich empfanden die zuständigen Stellen innerhalb der Stiftung dieses Bildzeichen als so sehr ihrer Sache angemessen, dass sie sich in den folgenden Jahren immer wieder auf Varianten dieses Plakats beschränkten: Geändert wurde lediglich die Farbstellung, insbesondere die Farbe des Bildgrundes. Nur ein einziges Mal, 1948, ist Pro Infirmis mit dem ausdrucksstarken, zeichnerisch meisterlichen Plakat von Hans Falk von dieser Plakatroutine abgewichen. Das Plakat mit der von einem Hund gehüteten, uns abgewandten jungen Frau im Rollstuhl ist vielen unvergesslich geblieben.
Bei dem wachsenden Anteil an Behinderten in unserer Bevölkerung und bei der zunehmenden Selbstverständlichkeit ihrer Integration ins Leben der Gemeinschaft sind vielleicht so aufrüttelnde Plakate wie jenes von Hans Falk nicht mehr notwendig. Ob in naher oder ferner Zukunft andere Plakatmotive oder -gestaltungsweisen auf die Anliegen von Pro Infirmis hinweisen werden, bleibt abzuwarten.

Pro Infirmis

Les œuvres humanitaires se distinguent non seulement par leurs activités et leurs objectifs – y compris les stratégies utilisées pour atteindre les buts fixés – mais encore par l'usage qu'elles font de l'affiche comme moyen publicitaire.
La Fondation Pro Infirmis, créée en 1920, s'est présentée au public en 1940 par une affiche. La main qui, dans un geste protecteur, porte délicatement une fleur gracieuse et fragile, est un motif qui traduit bien le sens de cette œuvre, tout en révélant les grandes qualités artistiques de l'affiche. Le projet, réalisé par Alois Carigiet, a bientôt été remplacé par une affiche de caractère symbolique de Donald Brun, exposée pour la première fois en 1944: elle montre une aile, de conception graphique soigneusement étudiée, qu'une chaîne empêche de se déployer pleinement.
Il faut croire que les organes compétents ont trouvé que ce symbole exprime si bien l'idée de la Fondation qu'ils se sont limités, les années suivantes, à publier des variations sur ce même thème, en modifiant simplement l'agencement des couleurs, en particulier la couleur du fond. Une seule fois, en 1948, Pro Infirmis s'est écartée de ce principe avec l'affiche de Hans Falk, exceptionnelle tant par sa force d'expression que par son dessin magistral. La jeune invalide qui, sous la garde de son chien, attend dans sa chaise roulante, est une image prenante et inoubliable.
Vu le nombre croissant de personnes handicapées dans notre population et leur intégration de plus en plus évidente à la vie de notre communauté, des affiches aussi bouleversantes que celle de Hans Falk ne sont plus nécessaires. Il est cependant difficile de savoir si, dans un avenir plus ou moins proche, d'autres motifs d'affiches ou formes de conception graphique susciteront à nouveau l'intérêt pour les activités de la Fondation Pro Infirmis.

154.

Alois Carigiet
1940, 90,5 × 128 cm
Lithography / Lithographie / Lithographie

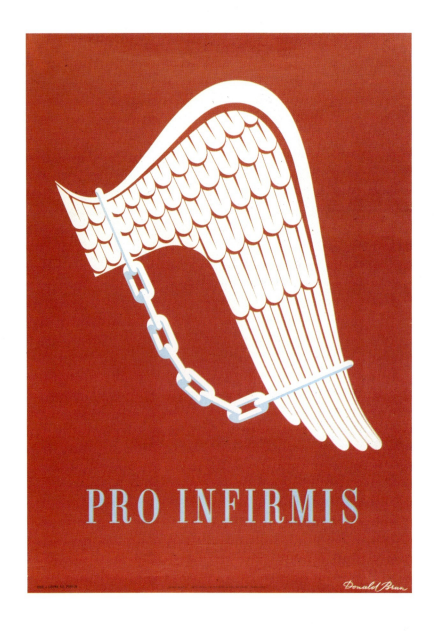

155.

Donald Brun
1944, 90,5 × 128 cm
Lithography / Lithographie / Lithographie

156.

Hans Falk
1948, 90,5 × 128 cm
Lithography / Lithographie / Lithographie

157.

Donald Brun
1970, 90,5 × 128 cm
Offset printing / Offsetdruck / Impression offset

158.

Robert Siebold
1983, 90,5 × 128 cm
Offset printing / Offsetdruck / Impression offset

Pro Juventute

Among the major social welfare organizations of Switzerland is the Pro Juventute foundation, dating from 1912, which in its work for young people is broadly equivalent to the Pro Senectute foundation, devoted to the care of old people. In the course of its history, Pro Juventute has operated with varying emphasis from the welfare of mother and infant to the problems of school-leavers and young adults (hostels and drug counselling, among others). Many of its activities are to be seen in the numerous posters issued by the foundation.

Starting in 1922 with a mother-and-child poster by Emil Cardinaux, and continuing into the Seventies, the Pro Juventute posters reflect half a century of Swiss poster history. While the poster style was at first rather conventional, stiff and given to sentimentality — possibly in accordance with the wishes of the client — a change becomes visible from 1930 with a poster by Otto Baumberger: the posters now show their subject from a fresh and often surprising side, with more than a little artistic boldness in the design of word and image.

From 1944/45 on, the Pro Juventute posters are among the best posters of each year. Well-known graphic artists participate: Alois Carigiet, Niklaus Stoecklin, Victor Rutz, Celestino Piatti and many others have helped to publicize Pro Juventute once or more. In general, a hand-drawn or painted style of poster is preferred. Surprisingly, a photographic approach has seldom been chosen, although the subject would seem almost to demand it. Perhaps the splendid, very painterly poster by Alois Carigiet has had a long-term effect as a "model" in this respect.

Pro Juventute

Unter den grossen Sozialwerken der Schweiz steht die der Jugend gewidmete, bereits 1912 gegründete Stiftung Pro Juventute gewissermassen der dem alten Menschen zugewandten Pro Senectute oder Stiftung für das Alter gegenüber. Von der Mütter- und Säuglingsfürsorge bis zu den Problemen der Schulentlassenen, der jungen Erwachsenen (Jugendhäuser, Drogenprobleme u. a.) hat sich Pro Juventute im Lauf ihrer Geschichte mit wechselnder Akzentsetzung beschäftigt. Viele ihrer Aktivitäten zeigen sich in den zahlreichen, von dieser Stiftung herausgegebenen Plakaten.

Einsetzend 1922 mit einem Mutter-Kind-Plakat von Emil Cardinaux, spiegeln, bis in die siebziger Jahre, die Pro-Juventute-Plakate über ein halbes Jahrhundert hinweg auch schweizerische Plakatgeschichte. War in den Anfängen der Plakatstil — vermutlich den Wünschen der Auftraggeber entsprechend — eher brav, bieder und nicht frei von Sentimentalität, so wird seit 1930 mit einem Plakat von Otto Baumberger ein Wechsel sichtbar: Die Plakate zeigen ihr Sujet nun von einer frischen, oft überraschenden Seite, in der Bild- und Schriftgestaltung werden in unterschiedlichem Mass plakatkünstlerische Kühnheiten gewagt.

Von 1944/45 an gehören die Plakate der Pro Juventute zu den Spitzen der jährlichen Plakatproduktion. Namhafte Graphiker sind daran beteiligt: Alois Carigiet, Niklaus Stoecklin, Victor Rutz, Celestino Piatti und manch andere haben ein- oder mehrmals für die Pro Juventute geworben. Bevorzugt wurde in der Regel ein zeichnerischer oder malerischer Plakatstil. Photographische Lösungen sind überraschenderweise selten gewählt worden, obwohl von der Thematik her sich solche Lösungen geradezu aufgedrängt hätten. Vielleicht hat in diesem Sinne das grossartige, sehr malerische Plakat von Alois Carigiet als ein «Exemplum» nachgewirkt.

Pro Juventute

La Fondation Pro Juventute, créée en 1912, est une œuvre d'aide sociale pour la jeunesse. Parmi les grandes institutions d'assistance en Suisse, elle constitue le pendant de la Fondation Pro Senectute pour la vieillesse. Son activité est marquée par des aspects aussi divers que la prévoyance maternelle et infantile, les problèmes des jeunes après la scolarité, à l'âge adulte (maisons de jeunesse, toxicomanie, etc.). Les nombreuses affiches publiées par la Fondation révèlent les priorités accordées à ces divers aspects tout au long de son histoire.

A commencer par une première affiche en 1922, avec le motif la mère et l'enfant d'Emile Cardinaux, jusque dans les années soixante-dix, les affiches Pro Juventute sont de précieux documents de l'histoire de l'affiche suisse pendant plus d'un demi-siècle. Si dans ses débuts — probablement pour répondre aux désirs des commettants — le style affichiste était plutôt sage, conventionnel, parfois même empreint de sentimentalité, un tournant s'est esquissé à partir de 1930 avec une affiche d'Otto Baumberger: les sujets, désormais d'une exquise fraîcheur, sont abordés sous un angle inattendu, la conception de l'image et la typographie font l'objet d'essais parfois audacieux.

A partir de 1944/45, les affiches Pro Juventute font chaque année partie des meilleures réalisations artistiques. Alois Carigiet, Niklaus Stoecklin, Victor Rutz, Celestino Piatti et bien d'autres graphistes connus ont créé une ou même plusieurs affiches pour contribuer à propager les activités de Pro Juventute. La préférence est en général donnée à un style fondé sur le dessin ou la peinture. Les interprétations photographiques ne sont que rarement envisagées, bien que du point de vue du thème, de telles solutions semblaient s'imposer. L'affiche grandiose et fort évocatrice d'Alois Carigiet a peut-être produit dans ce sens une orientation décisive.

159.

Emil Cardinaux
1922, 90,5 × 128 cm
Lithography / Lithographie / Lithographie

160.

Emil Cardinaux
1923, 90,5 × 128 cm
Lithography / Lithographie / Lithographie

161.

Emil Cardinaux
1926, 90,5 × 128 cm
Lithography / Lithographie / Lithographie

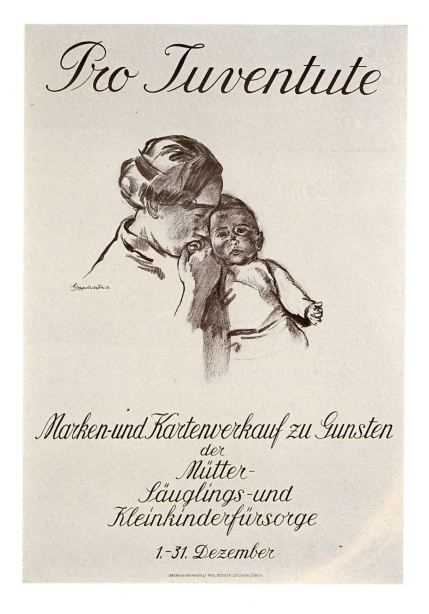

162.

Theodor Goppelsroeder
1927, 90,5 × 128 cm
Lithography / Lithographie / Lithographie

163.

Theodor Goppelsroeder
1928, 90,5 × 128 cm
Lithography / Lithographie / Lithographie

164.

Otto Baumberger
1930, 90,5 × 128 cm
Lithography / Lithographie / Lithographie

165.

Eric de Coulon
1932, 90,5 × 128 cm
Lithography / Lithographie / Lithographie

166.

Eric Hermès
1934, 90,5 × 128 cm
Lithography / Lithographie / Lithographie

167.

B. Kobi
1936, 70 × 100 cm
Lithography / Lithographie / Lithographie

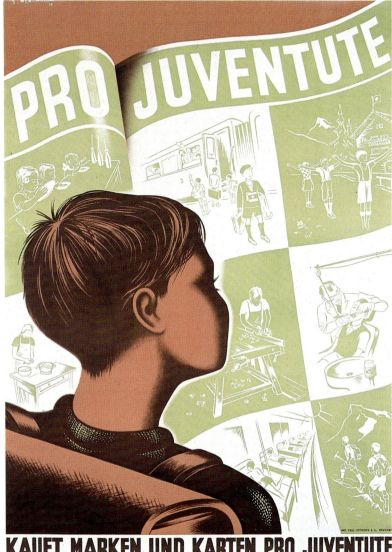

168.

Samuel Henchoz
1937, 90,5 × 128 cm
Lithography / Lithographie / Lithographie

169.

Samuel Henchoz
1938, 90,5 × 128 cm
Lithography / Lithographie / Lithographie

170.

Eugen Traugott Früh
1939, 90,5 × 128 cm
Lithography / Lithographie / Lithographie

171.

Eric Hermès
1940, 90,5 × 128 cm
Lithography / Lithographie / Lithographie

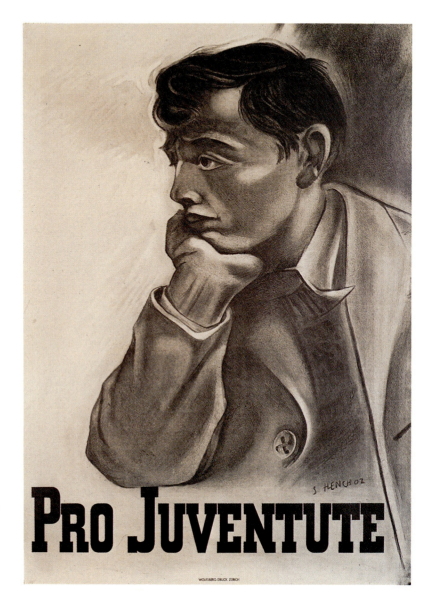

172.

Victor Rutz
1941, 90,5 × 128 cm
Lithography / Lithographie / Lithographie

173.

Samuel Henchoz
1942, 90,5 × 128 cm
Lithography / Lithographie / Lithographie

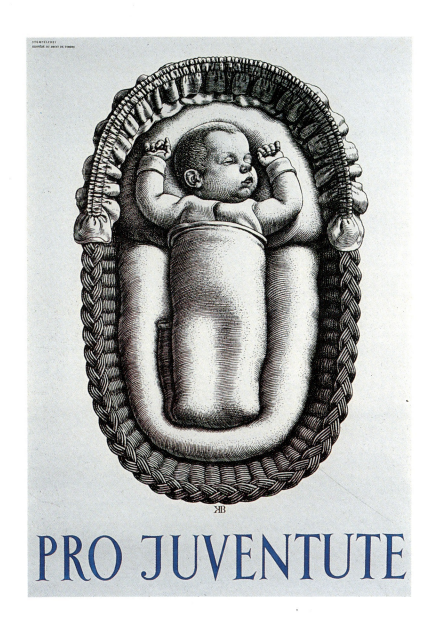

174.

Karl Bickel
1943, 90,5 × 128 cm
Lithography / Lithographie / Lithographie

175.

Rolf Bangerter
1944, 90,5 × 128 cm
Lithography / Lithographie / Lithographie

176.

P. Bouvier
1945, 90,5 × 128 cm
Lithography / Lithographie / Lithographie

177.

Friedrich Reinhard Brüderlin
1946, 90,5 × 128 cm
Lithography / Lithographie / Lithographie

178.

Alois Carigiet
1950, 90,5 × 128 cm
Offset printing / Offsetdruck / Impression offset

179.

Niklaus Stoecklin
1951, 70 × 100 cm
Lithography / Lithographie / Lithographie

180.

Victor Rutz
1952, 70×100 cm
Lithography / Lithographie / Lithographie

181.

Celestino Piatti
1955, 90,5×128 cm
Offset printing / Offsetdruck / Impression offset

182.

Celestino Piatti
1956, 90,5 × 128 cm
Lithography / Lithographie / Lithographie

183.

Celestino Piatti
1957, 90,5 × 128 cm
Lino cut / Linolschnitt / Gravure sur linoléum

184.

M. Lipps
1959, 70×100 cm
Offset printing / Offsetdruck / Impression offset

185.

Werner Christen
1960, 70×100 cm
Offset printing / Offsetdruck / Impression offset

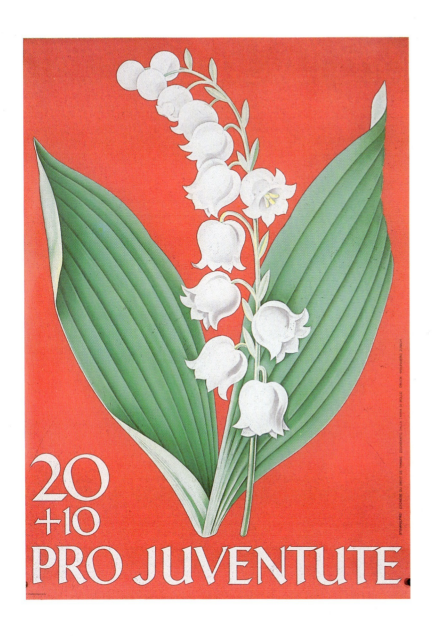
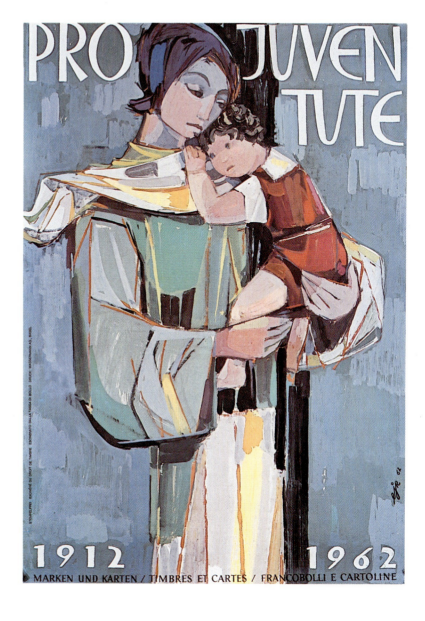

186.

Hans Schwarzenbach
1961, 70×100 cm
Lithography / Lithographie / Lithographie

187.

Eduard O. Renggli
1962, 70×100 cm
Offset printing / Offsetdruck / Impression offset

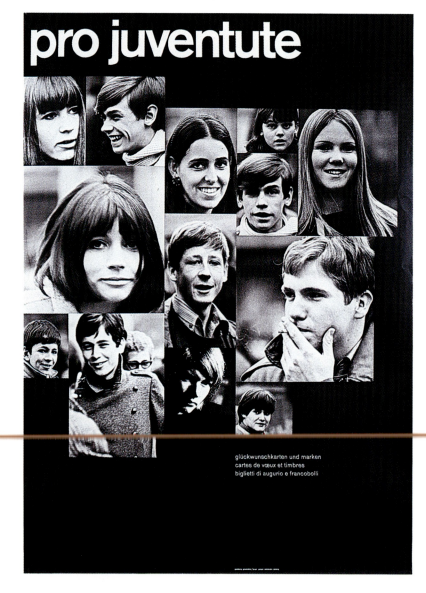

188.

Domenic Geissbühler
1966, 70×100 cm
Offset printing / Offsetdruck / Impression offset

189.

Domenic Geissbühler
1969, 90,5×128 cm
Silkprint / Siebdruck / Sérigraphie

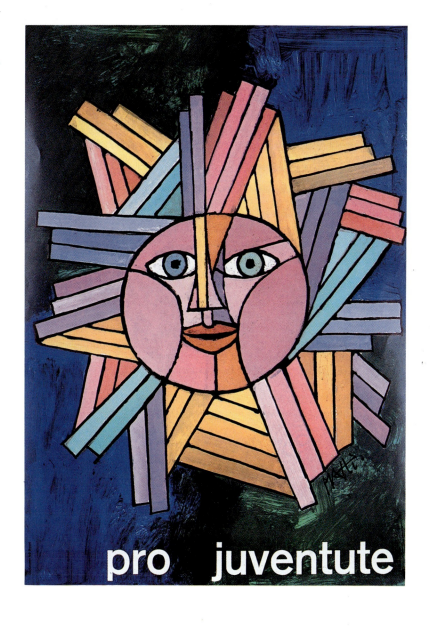

190.

Justesen / Atelier
1970, 90,5 × 128 cm
Offset printing / Offsetdruck / Impression offset

191.

Celestino Piatti
1972, 70 × 100 cm
Offset printing / Offsetdruck / Impression offset

Aid for Children

Since the beginning of the 20th century, a great variety of campaigns to provide aid for needy and handicapped children have been undertaken by organizations which have often been formed at short notice, sometimes also disappearing after a brief life. In the years of economic crisis between the two world wars, appeals on behalf of the children of the unemployed had particular importance. In this category of posters, the accusing drawing by the Basle painter Alfred Heinrich Pellegrini also indicates an artistically convincing position of social engagement. In contrast with this poster, which obtains its life entirely from the resources of draughtsmanship, an early example of the use of photography in poster art is hardly less effective. This is Gotthard Schuh's photo-poster of 1936, showing a young girl carrying her little brother and sister. The fact that she is taking the place of the mother gives the poster a particularly intensive human appeal.

The symbolic design of a "good luck" motif by the Zurich artist Ernst Keller has come to be regarded as a masterpiece of poster art. In support of a collection for the newly founded Pestalozzi children's village in 1946, he succeeded in establishing an identity between the badges offered for sale and the poster advertising them. Keller's response to the theme also shows how much more effective a stylized (and monumentalized) pictorial symbol is, in optical terms, than an illustrative representation of the theme, full of details.

Hilfsaktionen für das Kind

Von manchmal kurzfristig entstandenen, manchmal auch kurzlebigen Organisationen sind seit Anfang unseres Jahrhunderts vielerlei Hilfsaktionen zugunsten bedürftiger oder benachteiligter Kinder unternommen worden. Besonderes Gewicht hatten in den Krisenjahren zwischen den beiden Weltkriegen die Aktionen für die Kinder von Arbeitslosen.

In der Reihe der Plakate für solche Hilfsaktionen bezeichnet das anklägerische zeichnerische Plakat des Basler Malers Alfred Heinrich Pellegrini eine auch künstlerisch zwingende Position des sozialen Engagements. Diesem ganz aus den Möglichkeiten des Zeichnerischen lebenden Plakat antwortet ein völlig andersartiges, kaum weniger wirksames Plakat: Als ein frühes Beispiel der Nutzung der Photographie im Plakat darf das 1936 veröffentlichte Photoplakat von Gotthard Schuh gelten. Dass ein Mädchen sein kleines Geschwisterchen trägt, also gewissermassen an die Stelle der Mutter tritt, verleiht dem Plakat eine besonders intensive menschliche Note.

Zu einer Art Meisterplakat ist die signethafte Gestaltung des Glücksmotivs durch den Zürcher Ernst Keller geworden. Ihm gelang es 1946, bei einer Sammelaktion für das im Aufbau begriffene Kinderdorf Pestalozzi eine Identität zwischen dem zum Kauf angebotenen Abzeichen und dem dafür werbenden Plakat zu schaffen. Kellers Lösung zeigt überdies, um wie vieles ein stilisiertes (und monumentalisiertes) Bildsymbol optisch wirksamer ist als eine detailreiche, illustrative Darstellung des Themas.

Actions de secours en faveur de l'enfant

Depuis le début du XXe siècle, de nombreuses actions de secours en faveur d'enfants nécessiteux ou défavorisés ont été lancées par des organisations constituées parfois ad hoc ou à court terme. Pendant la crise économique entre les deux guerres mondiales, les actions de secours visaient en particulier à alléger le sort des enfants des chômeurs.

Dans la série d'affiches consacrées à de telles campagnes, le dessin vigoureux de l'affiche d'Alfred Heinrich Pellegrini exprime l'engagement social résolu du peintre bâlois, aussi sur le plan artistique. Tout aussi accusateur est le ton d'une affiche totalement différente, mais non moins percutante: l'affiche photographique de Gotthard Schuh, publiée en 1936, un des premiers exemples d'utilisation de la photographie dans une affiche. La fillette qui porte sa petite sœur, remplaçant ainsi en quelque sorte la mère, confère à l'affiche une note humaine particulièrement émouvante.

La conception stylisée du symbole du bonheur par le Zurichois Ernst Keller a abouti à une affiche véritablement magistrale. Pour une collecte organisée en 1946 en faveur du Fonds de construction du village d'enfants Pestalozzi, il a réussi à exprimer l'identité entre l'insigne proposé à la vente et l'affiche correspondante. La solution de Keller révèle d'ailleurs qu'un symbole visuel stylisé (et monumentalisé) a un plus grand impact qu'une présentation illustrative détaillée du thème.

192.

Carol Venalbes
1969, 53×77 cm
Offset printing / Offsetdruck / Impression offset

193.

Alfred Heinrich Pellegrini
1919, 90,5 × 127 cm
Lithography / Lithographie / Lithographie

194.

Carl Scherer
1927, 90,5 × 128 cm
Lithography / Lithographie / Lithographie

195.

Alois Carigiet
1933, 90,5 × 127 cm
Lithography / Lithographie / Lithographie

196.

Gotthard Schuh / Photo
1936, 90,5 × 128 cm
Intaglio / Tiefdruck / Impression en creux

197.

Paul Kammüller
1912, 46,5 × 68 cm
Lithography / Lithographie / Lithographie

198.

R. Weiss
1913, 73 × 105 cm
Lithography / Lithographie / Lithographie

199.

Oskar Weiss
1937, 70×100 cm
Lithography / Lithographie / Lithographie

200.

Ernst Keller
1946, 90,5×128 cm
Lino cut / Linolschnitt / Gravure sur linoléum

International Aid Appeals

Working groups with a specific humanitarian motive have often been formed in Switzerland in order to provide aid in cases of need which may have occurred unexpectedly. The aims and objects of such appeals, and particularly their conditions of operation, are not usually to be seen in a poster but are publicized through reports in the Press and the distribution of leaflets. The aid is usually for the benefit of civilian populations made destitute by acts of war.
Poster artists concerned with such humanitarian appeals generally offer their services free of charge. This may well bring the advantage that they are completely free in their work and have no obligation to take a client's views into account. Among the most impressive examples of such international aid posters are Heiri Steiner's powerfully drawn "Hungary" designs and Hans Erni's appeal on behalf of the civilian population of Vietnam.
Amnesty International has also repeatedly drawn attention to the plight of political prisoners, with posters in which photography has achieved a special documentary significance. This category of posters concerned with elementary human rights also includes a photo-montage by Werner Jeker, drawing renewed attention to the Declaration of the Rights of Man. One could well imagine that this very type of "political" poster might attain greater importance in a world which is becoming increasingly inhuman.

Internationale Hilfsaktionen

Immer wieder haben aus einer spezifisch humanitären Gesinnung heraus sich in der Schweiz Arbeitsgruppen gebildet, die sich für die Hilfe in einem vielleicht unerwartet aufgetretenen Notstand eingesetzt haben. Sinn und Zielsetzung solcher Aktionen, vor allem aber auch ihre Voraussetzungen, sind meist nicht aus einem Plakat ersichtlich, sondern waren von den unmittelbaren Zeitgenossen aus Pressekommentaren oder Flugblättern zu erfahren. Die Hilfsaktionen gelten meist der zivilen Bevölkerung, die durch Kriegshandlungen in Not geraten ist.
In der Regel stellen bei solchen humanitären Aktionen die Plakatgestalter ihre Entwürfe unentgeltlich zur Verfügung. Ihr Vorteil dabei mag sein, dass sie in der Gestaltung völlig frei sind und nicht auf die Argumente eines Auftraggebers Rücksicht nehmen müssen. Zu den stärksten Beispielen solcher Hilfsplakate gehören das zeichnerisch wirkungsvolle Ungarn-Plakat von Heiri Steiner und das Caritas-Plakat für die Zivilbevölkerung in Vietnam von Hans Erni.
Immer wieder hat auch die Organisation Amnesty International mit Plakaten auf das Los von politischen Gefangenen aufmerksam gemacht, wobei vielfach der Photographie besondere dokumentarische Bedeutung zukam. In die Reihe dieser an die elementaren Menschenrechte rührenden Plakate gehört auch die Photomontage von Werner Jeker zur Auffrischung der Erinnerung an die Declaration of the Rights of Man. Man könnte sich vorstellen, dass gerade dieser Typus von «politischem» Plakat in einer zunehmend unmenschlicher werdenden Welt zu vermehrter Bedeutung gelangen könnte.

Actions de secours internationales

Animés par une conception spécifiquement humanitaire, divers groupes de travail suisses se sont constitués spontanément pour offrir leur aide dans les situations d'urgence surgissant à l'improviste. Le sens et le but, et avant tout les multiples formes d'activité de ces groupes, ne sont guère exposés sur une affiche, mais sont diffusés par des communiqués de presse ou des tracts. Les actions de secours sont le plus souvent destinées à la population civile, durement affectée par les conflits entre les belligérants.
En règle générale, les affichistes se mettent gracieusement à la disposition de telles institutions humanitaires. Ils bénéficient en contrepartie d'une pleine liberté d'action dans leur travail créatif, sans devoir se soucier des arguments d'un commettant. Parmi les exemples les plus marquants, citons l'affiche sur la Hongrie de Heiri Steiner, impressionnante par la vigueur de son dessin, et l'affiche pour Caritas de Hans Erni, en faveur de la population civile au Viet-nam.
Le sort des prisonniers politiques a souvent été évoqué sur des affiches par l'organisation Amnesty International, la photographie prenant alors fréquemment une valeur documentaire. A cette série d'affiches rappelant les droits fondamentaux de l'homme appartient aussi le photomontage de Werner Jeker sur la Déclaration universelle des droits de l'homme. Ce type d'affiche «politique» pourrait fort bien gagner en importance dans un monde de plus en plus déshumanisé.

201.

Jules Courvoisier
1919, 90,5 × 128 cm
Lithography / Lithographie / Lithographie

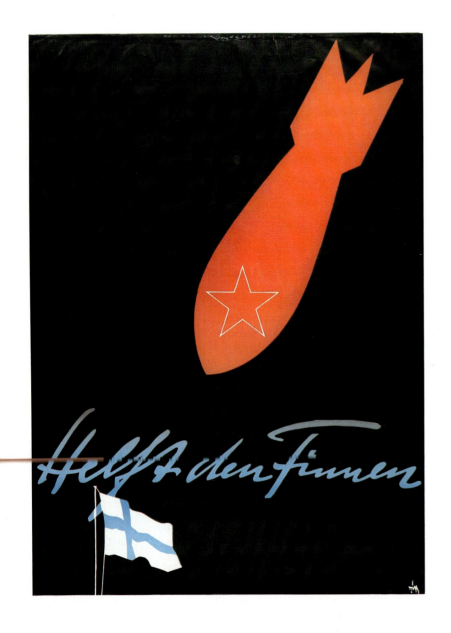

202.

Alex Walter Diggelmann
1941, 90,5×128 cm
Lino cut / Linolschnitt / Gravure sur linoléum

203.

Heiri Steiner
1956, 42×57,5 cm
Offset printing / Offsetdruck / Impression offset

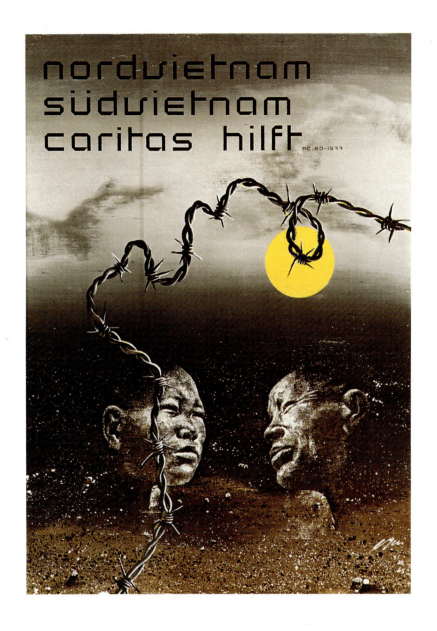
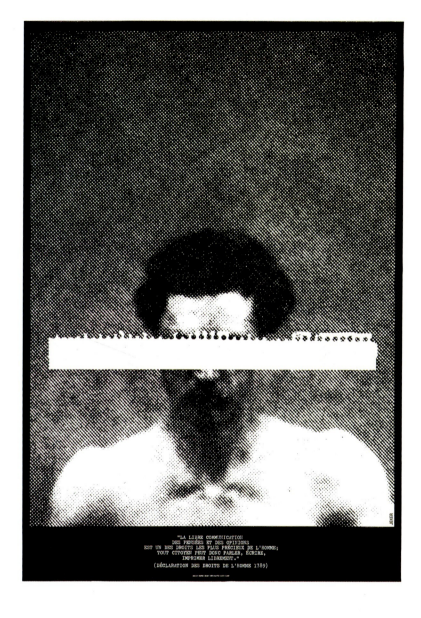

204.

Hans Erni
1968, 90,5 × 128 cm
Offset printing / Offsetdruck / Impression offset

205.

Werner Jeker
1982, 70 × 100 cm
Silkprint / Siebdruck / Sérigraphie

Protection of Nature and Wildlife

Naturschutz – Tierschutz

Protection de la nature – protection des animaux

The idea that Nature must be protected from the encroachments of mankind is certainly not a new one. As early as the beginning of this century, the Swiss Nature Protection League (Bund für Naturschutz) campaigned for new members with a poster. Since then, under constantly changing conditions, many campaigns have been undertaken for the protection of wild animals, plants, and occasionally whole regions. Domestic pets are also included in animal protection campaigns, as witnessed, for example, by the Niklaus Stoecklin poster of 1925 with its evocative cat's head. From time to time, the campaign against vivisection has also been included in this field.

In recent years, stimulated by heavy and sometimes almost irrevocable damage to the countryside, the protection of Nature has been extended to form an integral "environmental" theme. When Hans Erni drew attention to the need for the protection of water supplies in 1961, with a poster showing a death's head in startling combination with a glass of water, he provided a precursor to the theme of environmental protection. The same effect is made even more powerfully by the uncanny, surrealistic association of man and treetrunk in a 1983 poster by the same artist, conveying the message that the death of the forest eventually implies the death of mankind.

It is not impossible that increasing damage to the environment, combined with the strengthening of environmental consciousness, will lead to more and more poster campaigns designed to stimulate everyone to take part in the movement for survival.

Zwar ist der Gedanke, dass die Natur vor Übergriffen des Menschen geschützt werden muss, bereits ein alter Gedanke: Schon zu Anfang unseres Jahrhunderts ermunterte der Schweizerische Bund für Naturschutz mit einem Plakat die Bevölkerung zum Eintritt in diese Vereinigung. Seither wurden, unter stets wechselnden Voraussetzungen, viele Aktionen zum Schutz von Tieren und Pflanzen, gelegentlich von ganzen Landschaften unternommen. Beim Tierschutz – für den beispielsweise Niklaus Stoecklin 1925 mit dem suggestiven Katzenkopf warb – spielen auch Probleme der Haustierhaltung eine Rolle. Und zu gewissen Zeiten wurde auch ein in diesen Rahmen gehörender Kampf gegen die Vivisektion geführt.

In den letzten Jahren hat sich, wachgerüttelt durch schwere, teils bereits kaum mehr behebbare Schäden, der Naturschutz-Gedanke ausgeweitet zu einem integralen «Umweltschutz». Wenn Hans Erni bereits 1961 mit dem aufrüttelnden Totenkopf über dem Wasserglas auf die Notwendigkeit der Rettung des Wassers hingewiesen hat, so ist dies bereits ein Vorläufer der Umweltschutz-Thematik. In noch stärkerem Masse gilt dies für ein Plakat, das 1983 unübersehbar auf das Waldsterben aufmerksam machte. In einer surreal-unheimlichen Verbindung von Baumstrunk und Mensch veranschaulicht Hans Erni, dass der Tod des Waldes letztlich zugleich unseren Tod bedeutet.

Es ist nicht ausgeschlossen, dass die wachsenden Umweltschäden einerseits, ein erstarkendes Umweltbewusstsein anderseits in den kommenden Jahren vermehrt zu Plakataktionen führen, mit denen jedermann aufgerüttelt werden soll, sein Teil am Überleben zu leisten.

L'idée de protéger la nature contre l'emprise de l'homme n'est pas nouvelle: déjà au début de ce siècle, la Ligue suisse pour la protection de la nature invitait la population à adhérer à cette association. Depuis lors, face à des conditions souvent changeantes, d'innombrables campagnes ont été lancées pour protéger les animaux et les plantes, parfois même le paysage de toute une région. La protection des animaux – que Niklaus Stoecklin a, par exemple, illustré en 1925 par une tête de chat suggestive – inclut aussi des problèmes relatifs aux animaux domestiques. A certaines époques, la lutte contre la vivisection a également été menée par voie d'affiches.

Ebranlés par les graves dommages, en partie irréversibles, survenus au cours de ces dernières années, les protagonistes de la protection de la nature ont élargi l'idée en une protection intégrale de l'environnement. L'affiche avec une tête de mort pathétique au-dessus d'un verre d'eau, par laquelle Hans Erni attirait déjà en 1961 l'attention sur la nécessité de sauver l'eau, est un précurseur du thème de la protection de l'environnement. Plus dramatique encore est l'affiche qui signale en 1983 le dépérissement de la forêt: dans une association surréaliste angoissante entre un tronc d'arbre et un homme, Hans Erni illustre l'idée que la mort de la forêt est finalement aussi notre mort.

Il n'est pas exclus que les dégradations croissantes de notre environnement, d'une part, et une conscience écologique plus sensibilisée, d'autre part, ne suscitent au cours de ces prochaines années des campagnes plus nombreuses, visant à rappeler par des affiches suggestives que chacun est appelé à fournir sa contribution à la survie de notre environnement.

206.

Anton Christoffel
1908, 63 × 93 cm
Lithography / Lithographie / Lithographie

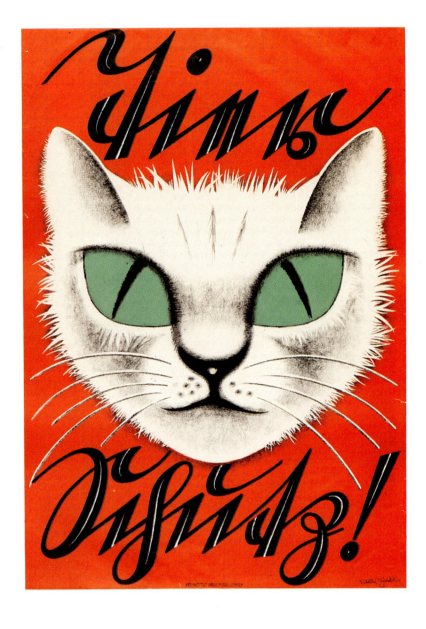

207.

H. Gattiker
1921, 45×64 cm
Lithography / Lithographie / Lithographie

208.

Niklaus Stoecklin
1925, 90,5×128 cm
Lithography / Lithographie / Lithographie

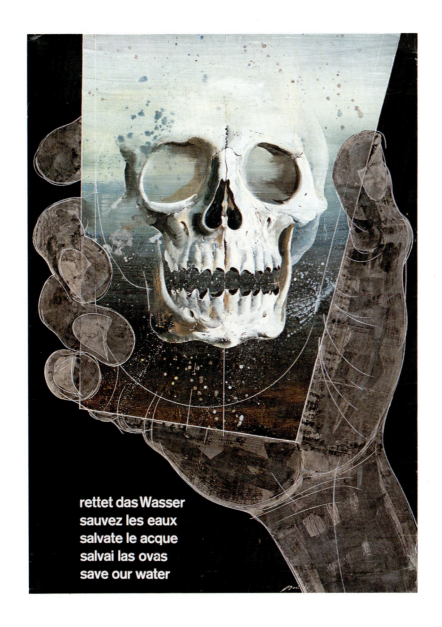
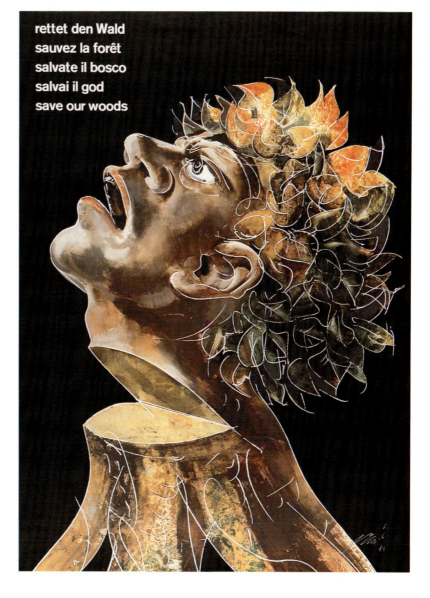

209.

Hans Erni
1961, 90,5 × 128 cm
Intaglio / Tiefdruck / Impression en creux

210.

Hans Erni
1983, 90,5 × 128 cm
Offset printing / Offsetdruck / Impression offset

Protection of the Environment

The problems of protecting our environment from exploitation and destruction have become everyday matters, in the sense that they can no longer be evaded or ignored by anyone. Instinctively or through understanding we are all aware that we have reached the frontiers of growth, and we are already more or less accustomed to reading bad news on the subject in the papers every day.
It is not only a matter of the major problems of survival raised by the destruction of the resources necessary for life itself. There is also, for example, the matter of noise from car and motor-cycle traffic on city streets and motorways. As long ago as 1960, an effective photo-poster by Josef Müller-Brockmann campaigned for protection from a degree of noise which is both physically and mentally unbearable for many people.
There is also the question of the rapidly growing "rubbish mountain", which we all help to produce and are now barely able to eliminate without causing new forms of damage. On this theme, attention was created in 1971 by an Edgar Küng poster featuring sacks full of rubbish. There is also increasing publicity for recycling processes, for glass and paper, for example. Poster campaigns for the re-use of other products can be expected. A high standard of poster quality can be useful in this respect, proclaiming a message effectively and making it almost impossible to evade. The poster catches the eye of everyone who passes by.

Die Umwelt schützen – uns vor Immissionen schützen

Probleme der Umweltbelastung, der Umweltzerstörung oder des Umweltschutzes sind zu Alltagsproblemen geworden, in dem Sinne nämlich, dass keiner ihnen mehr ausweichen kann, keiner sich ihnen mehr verschliessen kann. Instinktmässig oder verstandesmässig erleben und erfahren wir alle, dass wir an den Grenzen des Wachstums angelangt sind. Und beinahe ist man es schon gewohnt, dass die Morgenzeitung eine neue Hiobsbotschaft bringt.
Dabei geht es nicht nur um die sogenannt grossen Probleme des Überlebens überhaupt, das heisst die Zerstörung der für uns lebensnotwendigen Ressourcen. Es geht zum Beispiel auch um den Lärm (des Auto- und Motorradverkehrs auf städtischen Strassen wie auf Autobahnen). Um den Schutz vor dem für viele Menschen physisch wie psychisch unerträglichen Lärm hat schon 1960 ein wirkungsvolles Photoplakat von Josef Müller-Brockmann gebeten.
Es geht aber auch um den rasch wachsenden Abfall- und Kehrichtberg, den wir alle produzieren und kaum mehr zu beseitigen imstande sind, ohne neue Schäden zu verursachen. Auf diese Thematik hat 1971 das Plakat von Edgar Küng mit den uns anblickenden Kehrichtsäcken aufmerksam gemacht. Vermehrt gibt es auch eine Werbung für das Recycling, etwa von Glas oder Papier. Plakataktionen für die Wiederverwertung auch anderer Produkte sind zu erwarten. Dabei kann eine bestimmte Qualität des Plakats nützlich sein: Es verkündet seine Botschaft sehr wirkungsvoll, man kann ihm fast nicht ausweichen. Und: es «erwischt» jeden, der sich im öffentlichen Raum bewegt.

Protéger l'environnement – se protéger contre les immissions

La pollution et la dégradation de l'environnement, et partant la protection de notre milieu naturel, sont des problèmes auxquels nous sommes confrontés quotidiennement en ce sens que nul ne peut plus y échapper ou prétendre les ignorer. Nos instincts ou notre raison nous apprennent que nous avons atteint les limites de la croissance. Et déjà nous avons presque pris l'habitude de voir chaque matin les journaux annoncer une nouvelle catastrophe.
Il ne s'agit toutefois pas seulement des soi-disant grands problèmes de la survie en général, c'est-à-dire de la destruction de nos ressources vitales, mais encore des multiples problèmes journaliers, tels que les immissions (du trafic motorisé sur le réseau urbain et sur les autoroutes). La protection contre le bruit, physiquement et psychiquement insupportable pour tant d'hommes, a déjà fait l'objet d'une affiche photographique suggestive de Josef Müller-Brockmann en 1960.
Il y a ensuite la masse grandissante des déchets et ordures que nous produisons et que nous n'arrivons presque plus à éliminer, sans provoquer de nouveaux dommages. Ce sujet a été traité en 1971 par Edgar Küng, sous forme de sacs à ordures qui nous regardent. La propagande pour le recyclage, par exemple de verre ou de papier, gagne en importance sur les murs d'affichage. Des campagnes pour la récupération d'autres produits encore sont concevables. La qualité de l'affiche peut alors devenir déterminante: le message est diffusé avec efficacité, nul ne peut pratiquement s'y soustraire. Tout le monde en prend connaissance, chacun se sent «concerné».

211.

Josef Müller-Brockmann
1960, 90,5 × 128 cm
Offset printing / Offsetdruck / Impression offset

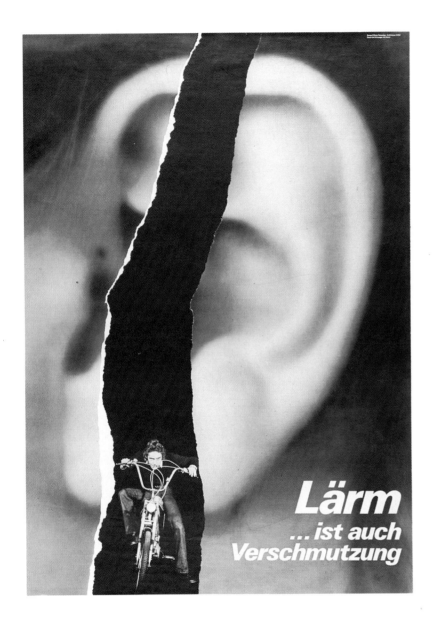
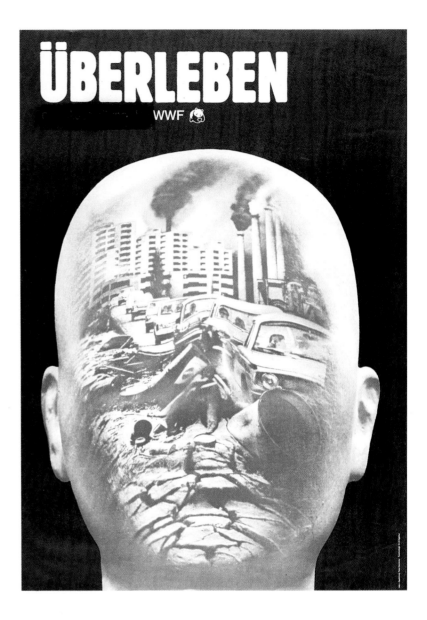

212.

Olivier Delacrétaz
1967, 90,5 × 128 cm
Offset printing / Offsetdruck / Impression offset

213.

Peter Hajnoczky
1970, 90,5 × 128 cm
Offset printing / Offsetdruck / Impression offset

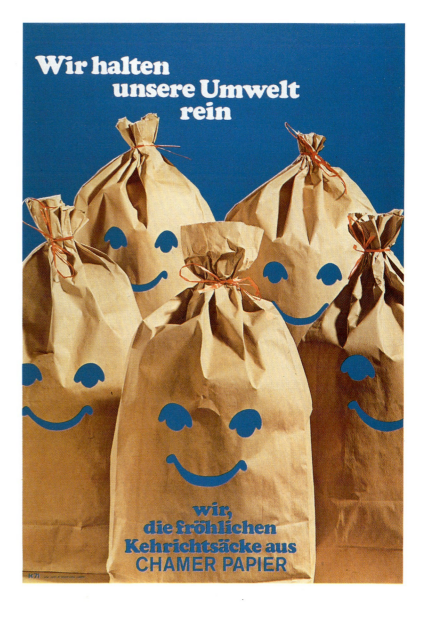

214.

Heiner Jenny
1971, 90,5 × 128 cm
Offset printing / Offsetdruck / Impression offset

215.

Edgar Küng
1971, 90,5 × 128 cm
Offset printing / Offsetdruck / Impression offset

Campaign against Drug Abuse

Posters issued by city or cantonal police authorities have become increasingly important in campaigns to warn the population against certain dangers. Most conspicuous are posters warning against pickpockets, muggers and burglars of holiday homes out of season. The increasing risks to women in little-frequented and badly-lit districts at night have also given rise to poster campaigns. As early as 1972, the Zurich City Police drew attention to the dangers facing women going home alone at night through deserted streets, with an effective visualization of the slogan: "Dark street — dark purpose" ("Dunkler Weg — dunkle Absicht").

It is a striking fact that black-and-white photography or photo-montage is generally used, very often with dramatic effect, in posters warning against the dangers of accidents and crime. This is doubtless due to a conviction that a particularly powerful impression is made on the viewer by a realistic (or seemingly realistic) photographic evocation of a dangerous situation.

A macabre photo-montage, which many people at the time found excessively shocking, was the subject of a poster issued by the Zurich City Police in 1969 as the first in a campaign about the increasing danger of drug addiction among young people. Andreas Fierz, the designer, superimposed a mask-like skull and a vertebral column on the half-length portrait of a young woman. The strong effect of the poster is based on the merging of life with death, while the viewer feels personally and directly addressed by the staring eyes which look warningly (or piteously?) through the huge sockets of the skull.

In warnings against dangers and crime, and particularly in the campaign against drug addiction, the poster gains a new kind of meaning within contemporary society: it takes over the role of "helping hand". Perhaps it manages this job especially well and effectively because, in the anonymity of the street hoarding, it does not insistently moralize but mutely, yet eloquently, addresses everyone. The viewers are left entirely free to accept or reject the constantly repeated message.

Kampf gegen Rauschgift

Im Rahmen der Plakataktionen zur Warnung der Bevölkerung vor bestimmten Gefahren haben die von städtischen oder kantonalen Polizeiämtern herausgegebenen Plakate zunehmende Bedeutung erhalten. Im Vordergrund steht die Warnung vor Taschendieben, vor Entreiss-Diebstählen oder Einbrüchen in ferienzeitlich unbewohnte Wohnungen. Auch die wachsende nächtliche Gefährdung von Frauen in wenig begangenen oder schlecht beleuchteten Gegenden hat zu Plakataktionen geführt. So hat etwa schon 1972 die Stadtpolizei Zürich mit einer effektvollen Visualisierung des Slogans «Dunkler Weg – dunkle Absicht» auf die Gefahren aufmerksam gemacht, denen sich eine Frau aussetzt, wenn sie nachts allein durch einsame Strassen nach Hause geht.

Auffallenderweise wird für Warnplakate, die mit Gefährdungen durch Unfälle oder Verbrechen zu tun haben, sehr oft die dramatisch wirkungsvolle, meist schwarzweisse Photographie oder Photomontage eingesetzt. Dies zweifellos aus der Überzeugung heraus, dass von der realistischen (oder scheinrealistischen) photographischen Evokation einer gefährlichen Situation eine besonders starke Wirkung auf den Betrachter ausgeht.

Mit einer – damals von vielen als fast unerträglicher Schock empfundenen – makabren Photomontage hat 1969 das von der Stadtpolizei Zürich herausgegebene Plakat «Rauschgift» die zunehmende Gefährdung Jugendlicher durch die Drogensucht erstmals zum Plakatthema gemacht. Andreas Fierz hat ins Brustbild einer jungen Frau wie eine Maske einen Totenschädel eingesetzt und in den Hals die Wirbelsäule eines Skeletts montiert. Die starke Wirkung des Plakats beruht auf der Durchdringung von Leben und Tod, wobei sich der Betrachter persönlich und direkt durch das starre Augenpaar angesprochen fühlt, das aus den riesigen Augenhöhlen des Totenkopfs hervor ihn warnend (oder hilfesuchend?) anblickt.

Bei der Warnung vor Gefahren und Verbrechen, vor allem aber im Kampf um das Rauschgift erhält das Plakat eine neuartige Bedeutung innerhalb der zeitgenössischen Gesellschaft: Es übernimmt die Funktionen einer helfenden Hand. Und es bewältigt diese Aufgabe vielleicht besonders gut und wirksam, weil es in der Anonymität des Strassenaushangs seine «Moral» nicht penetrant aufschwätzt, sondern jeden stumm und zugleich beredt anspricht. Dem auf diese Weise Angesprochenen ist die volle Freiheit belassen, ob er die wieder und wieder geäusserte Botschaft annehmen will oder nicht.

La lutte contre la drogue

Dans le cadre des campagnes destinées à avertir la population des dangers qui la menacent, les affiches publiées par les services de la police municipale ou cantonale prennent une importance croissante. La mise en garde contre les pickpockets, le vol d'objets arrachés par un acte de violence ou les cambriolages de logements de vacances inoccupés prennent une place de premier ordre. Les dangers qui menacent de plus en plus souvent les femmes dans les régions peu fréquentées ou mal éclairées ont également suscité des séries d'affiches. Sous forme de visualisation suggestive du slogan «Chemin sombre – intentions obscures», la police de la ville de Zurich a attiré dès 1972 l'attention sur les dangers qui guettent une femme rentrant seule la nuit.

Il est frappant de constater que pour les affiches de mise en garde, signalant les risques dus à des accidents ou des crimes, les artistes recourent très souvent à la photographie ou au photomontage noir/blanc, produisant un effet dramatique intense. L'évocation photographique réaliste (ou pseudo-réaliste) d'une situation dangereuse exerce incontestablement un puissant impact sur l'observateur. L'affiche sur la drogue, publiée en 1969 par la police de la ville de Zurich sous forme de photomontage macabre, a été ressentie par beaucoup comme un choc quasiment insupportable. La toxicomanie de plus en plus fréquente chez les jeunes était ainsi devenue pour la première fois le thème d'une affiche. Sur le thorax d'une jeune femme, Andreas Fierz a monté une tête de mort à la manière d'un masque, et dans le cou il a inséré la colonne vertébrale d'un squelette. L'expressivité de l'affiche découle de l'interpénétration profonde entre la vie et la mort. L'observateur se sent sollicité, personnellement et directement, par les yeux figés qui, du fond des gigantesques orbites de la tête de mort, lui lancent un avertissement (ou implorent son aide?).

Signal expressif dans la lutte contre les dangers et les crimes, et plus spécialement contre la toxicomanie, l'affiche acquiert une nouvelle fonction dans la société moderne: celle de la main secourable. Elle assume ce rôle avec d'autant plus de persuasion qu'elle n'impose pas sa «morale» avec insistance. Perdue dans l'anonymat de l'affichage public, elle aborde le passant sans parler, mais avec d'autant plus d'éloquence, laissant à chacun l'entière liberté d'accepter ou de refuser le message constamment répété.

216.

Andreas Fierz
1969, 90,5 × 128 cm
Offset printing / Offsetdruck / Impression offset

217.

Bruno Teucher
1983, 70×100 cm
Offset printing / Offsetdruck / Impression offset

Artist	**Künstler**	**Artiste**

Aeschbach, Hans (1911–) 130
Afflerbach, Beatrice 25, 26
Andermatt, Peter (1938–) 104, 105

Bäder, Heinz 113
Bangerter, Rolf (1922–) 175
Bänninger, Urs 79
Bauer, Heiner (1922–1981) 54
Baumann, Ruedi 78
Baumberger, Otto (1889–1961) 14, 125, 141, 142, 143, 144, 145, 164
Barth, Ruedi 134
Behrmann/Bosshard 90
Bickel, Karl (1886–1982) 1, 174
Bouvier, P. 176
Brauchli, Pierre 74
Brüderlin, Friedrich Reinhard 177
Brun, Donald (1909–) 22, 33, 155, 157
Buchdruckerei Baden AG 39

Cardinaux, Emil (1888–1956) 159, 160, 161
Carigiet, Alois (1902–) 154, 178, 195
Cathomas, Andreas 108
Christen, Werner 185
Christoffel, Anton 206
Coulon, Eric de (1888–1956) 165
Courvoisier, Jules (1884–1936) 48, 116, 201

Däppen, Erwin 57
Delacrétaz, Olivier 212
Diggelmann, Alex Walter (1902–) 202
Duss, Carlos 84

Erni, Hans (1909–) 7, 13, 23, 64, 204, 209, 210

Falk, Hans (1918–) 128, 131, 138, 139, 140, 156
Fässler, Franz (1920–) 93
Fierz, Andreas 216
Fontanet, Noël (1898–1982) 28
Freis, Peter 41
Früh, Eugen 147, 170

Gallay, Michel 151
Galli, Orio 80
Gams, Margrit 15
Gantert, Hans 69, 85
Gattiker, H. 207
Gauchat, Pierre (1902–1956) 117
Geissbühler, Domenic (1932–) 188, 189
Gerber, E. 87
Gilsi, René (1905–) 35, 55
Goppelsroeder, Theodor 162, 163
Guignard, Roland 91

Hajnoczky, Peter (1943–) 107, 213
Handschin, Hans (1899–1948) 126

Hangarter, Basil 114
Hauth, Dora (1874–1957) 16, 17, 52
Henchoz, Samuel (1905–1976) 168, 169, 173
Hermès, Eric 166, 171
His, Andreas 102
Honegger, Gottfried 86
Hotz, Emil 98, 101
Hürzeler, Peter 66

Jacopin, Paul 92
Jeker, Werner (1944–) 214
Jenny, Heiner 214
Justesen, Atelier 190

Kammüller, Paul (1885–1946) 43, 46, 197
Keiser, Ernst 20
Keller, Ernst (1891–1968) 97, 129, 148, 149, 200
Kobi, B. 167
König, Peter 76
Kramer, Pierre 120, 123
Küng, Edgar 153, 215

Lämmler, Jakob 88
Laubi, Hugo (1888–1959)
Lipps, M. 184

Masmejan, André (1932–) 150, 152
Miedinger, Gérard (1914–) 95, 96, 136, 137
Moll, Florentin (1884–1942) 51
Monnerat, Pierre (1917–) 94
Monticelli, Walter 10
Müller-Brockmann, Josef (1914–) 211
Müller, Fridolin (1926–) 135

Neuburg, Hans (1904–1983) 146

Peikert, Martin (1901–1975) 127
Pellegrini, Alfred Heinrich (1881–1958) 2, 18, 50, 193
Pfalzburger, Ruth 110
Piatti, Celestino (1922–) 32, 133, 181, 182, 183, 191

Renggli, Eduard (1882–1939) 187
Rutz, Victor (1913–) 172, 180

Scherer, Carl (1890–1953) 3, 5, 6, 44, 45, 81, 82, 194
Schierle, Ulrich 106
Schlatter, Ernst Emil (1883–1954) 124
Schlup, Bernard (1948–) 58, 71
Schmid, W. 152
Schönhaus, Cioma 65
Schuh, Gotthard 196
Schuhmacher, Hugo (1939–) 70, 77, 83
Schwarzenbach, Hans 196
Siebold, Robert 158

Sollberger, Paul 99, 100
Spahr, Jürg 31
Staub, R. 12
Stauffer, Jürg 73
Steinberger, Emil 109
Steiner, Heiri (1906–1983) 203
Stoecklin, Niklaus (1896–1983) 19, 179, 208
Sutter, Erika 72

Teucher, Bruno 220

Venalbes, Carol 192
Vivarelli, Carlo (1919–) 132

Wälti R. 67
Wehrli, 103
Weiss, E. 118
Weiss, Oskar 199
Weiss, R. 198
Wenk, Wilhelm 29
Wieland, Hans Beat (1867–1945) 4, 115, 119
Wild, Walter 112
Wolff, Steffen 111
Wyss, Paul (1875–1952) 47

Printer | **Drucker** | **Imprimeur**

Atar SA, Neuchâtel   67, 120, 123, 165, 169

Bender, Paul, Zürich   100, 107, 163, 172, 175, 176
Berichthaus, Zürich   97, 200
Bollmann, J., Zürich   101, 183
Bubenberg-Druck, Basel   71, 85
Buchdruckerei Baden AG   39
Bucher, C. J., Luzern   215
Bösch, Luzern   153

Casserini & Sohn, Thun   179, 180
Cedips, Lausanne   61
City-Druck, Zürich   91
Conzett + Huber, Zürich   104, 209

Dickmann AG, Basel   191
Du Bourg, Martigny   72
Du Pré-Jérôme, Genève   34

Eidenbenz & Co., St. Gallen   109, 112

Fiedler SA, La Chaux-de-Fonds   27
Fretz, Gebr., AG, Zürich   3, 4, 5, 21, 49, 90, 115, 119, 122, 124, 198
Frey-Wiederkehr, Zürich   54
Frobenius AG, Basel   182

Gamper, Rümlang   189
Geisseler-Druck, Zürich   73
Gemsberg, Basel   50
Gerber, Peter, Zürich   84
Graphische Betriebe Coop, Basel   13, 86

Hertig + Co. AG, Biel   40
Hofer & Co., Zürich   16, 17, 52
Hug + Söhne AG, Zürich   37, 105, 114, 213
Hutter AG, Wohlen   58

Kümmerly & Frey AG, Bern   96, 99, 103, 111, 126, 134

Lichtdruck AG, Dielsdorf   42
Lienberger, Zürich   214
Löpfe-Benz AG, Rorschach   158

Marsens, Lausanne   95, 151

Maurer, Gebr., Zürich   8, 9, 10
Maurer, Emil G., Zürich   35, 55
Morf, J. J., Basel   20, 43
Müller, J. C., AG, Zürich   24, 82, 93, 108, 121, 125, 128, 129, 136, 137, 138, 139, 140, 141, 142, 143, 144, 145, 148, 149, 156, 164, 167, 190, 199, 216

Nationalzeitung, Basel   65

Orell Füssli AG, Zürich   15, 130, 132, 196, 208

Pfister AG, Winterthur   38

Schwegler, Karl, AG, Zürich   212
Sonor SA, Genève   48, 116, 201
Säuberlin & Pfeiffer SA, Vevey   166, 171, 185

Trüb & Cie, Aarau   12, 41

Uldry, Albin, Bern   205

Vogt, Edwin, Waldenburg   11, 66, 69, 70, 74, 76, 77, 78, 79, 83

Wassermann AG, Basel   2, 19, 22, 23, 26, 29, 30, 31, 32, 33, 36, 46, 47, 57, 98, 102, 106, 110, 113, 133, 155, 181, 187, 188, 197
Wolf AG, Basel   193
Wolfensberger, J. E., AG, Zürich   7, 14, 92, 94, 131, 147, 154, 159, 160, 161, 162, 170, 173, 177, 178, 184, 186, 195, 207, 210,

Ziltener, A., Winterthur   217

Bildnachweis

Plakatsammlung Kunstgewerbemuseum Zürich
Nr. 158 Pro Infirmis Zürich
Nr. 1   Landesbibliothek Bern

Kein Teil dieses Buches darf ohne schriftliche Genehmigung des Verlages in irgendeiner Form durch Photokopie, Mikrofilme oder andere Verfahren reproduziert werden.

Format
Weltformat = 90,5 × 128 cm, Breite × Höhe

Impressum

Authors
Willy Rotzler
Karl Wobmann

Translations
Andrew Bluhm, English
Denise Schai, French

Graphic conception
Henry D. Béguelin

Book-jacket
Henry D. Béguelin

Responsible for publication
Konrad Baumann

Color separations
Cliché+Litho AG, Zurich

Total production
Offset+Buchdruck AG, Zurich

Impressum

Autoren
Willy Rotzler
Karl Wobmann

Übersetzungen
Andrew Bluhm, Englisch
Denise Schai, Französisch

Grafische Gestaltung
Henry D. Béguelin

Schutzumschlag
Henry D. Béguelin

Verlegerische Gesamtleitung
Konrad Baumann

Fotolithos
Cliché+Litho AG, Zürich

Gesamtherstellung
Offset+Buchdruck AG, Zürich

Impressum

Auteurs
Willy Rotzler
Karl Wobmann

Traductions
Andrew Bluhm, anglais
Denise Schai, français

Conception graphique
Henry D. Béguelin

Couverture du livre
Henry D. Béguelin

Production générale
Konrad Baumann

Photolithographies
Cliché+Litho AG, Zurich

Production générale
Offset+Buchdruck AG, Zurich